Magic Multiplication

Discover the Ultimate Formula for Fast Multiplication

Professor Chengqi ZHANG

Magic Multiplication

Discover the Ultimate Formula for Fast Multiplication

Professor Chengqi ZHANG

ISBN-13: 978-988-8843-27-5

© 2023 Chengqi ZHANG

Edited by Adeline Sun

MATHEMATICS / General

EB196

All rights reserved. No part of this book may be reproduced in material form, by any means, whether graphic, electronic, mechanical or other, including photocopying or information storage, in whole or in part. May not be used to prepare other publications without written permission from the publisher except in the case of brief quotations embodied in critical articles or reviews. For information contact info@earnshawbooks.com

Published by Earnshaw Books Ltd. (Hong Kong)

FOREWORD

65 x 65 = 4225   56 x 56 = 3136   47 x 47 = 2209

108 x 108 = 11664   94 x 94 = 8836   72 x 78 = 5616

27 x 87 = 2349   73 x 87 = 6351   53 x 65 = 3445

106 x 109 = 11554   92 x 97 = 8924...

Can you work out the results of these calculations without a calculator or pen and paper? Of course you can, and so can anyone else who reads this book and learns the tricks it contains. All you need to be able to do is calculation of the multiplication of any two-digit numbers! This book introduces several magical rapid calculation formulas with examples and easy-to-understand methods. It teaches you to embrace your innate mathematical talent and takes you into the world of magic numbers.

The primary readers of this book are middle school students, but it works for anyone. Do you worry about your math capabilities? Try this book and learn some magical rapid calculation formulas, then quickly solve problems such as 59 x 59 =? and 46 x 54 =?

The focus of this book is not only to teach rapid calculation formulas, but also to stimulate enthusiasm for math through rapid calculation, to cultivate confidence in calculation, and to understand the essence of calculation rules, so as to lay a solid mathematical foundation for further learning. Learning rapid calculation can save a lot of homework time and this book will stimulate your child's interest in mathematics.

The goal is to express profound things in simple language, and to inspire readers to discover the laws governing numbers, and appreciate the magic and mystery of calculation so as to improve their self-learning ability. No esoteric math theory or training is required. As long as you can multiply two-digit numbers, add and subtract, and understand simple negative numbers, you can learn the basics of this book.

There are only three important formulas introduced, and it only takes an average of about ten hours for middle school students to learn the basic algorithms, allowing you to mentally calculate the square of all numbers up to 125. After about ten more hours working through the exercises, you will be proficient in all the algorithms. With another ten or so hours, you'll be an expert in magic multiplication.

If you're a high school graduate who likes arithmetic, you can skip directly to the proofs in Chapters 5 and 8 and learn all the rapid calculations in a few hours.

Give it a try! Let's explore the beauty of mathematics together!

TABLE OF CONTENTS

Foreword	3
How to Use This Book	9
Self-Assessment Questions	11
Part I: Magic Square Speed Algorithm	13
Chapter 1: Introduction	15
Section 1.1: Background Knowledge – Nine-by-Nine Multiplication Table (difficulty 1*)	15
Section 1.2: Fun Facts – Square of Numbers Whose Ones Digit is 5 (difficulty 2*)	15
Chapter 2: Square Numbers Without Carrying	19
Section 2.1: 51 – 59 Squared (difficulty 2*)	20
Section 2.2: 41 – 49 Squared (difficulty 3*)	22
Section 2.3: 101 – 109 Squared (difficulty 2*)	25
Section 2.4: 91 – 99 Squared (difficulty 3*)	27
Section 2.5: Summary (difficulty 3*)	29
Chapter 3: Square Numbers with Carrying	31
Section 3.1: 11 – 24 Squared (difficulty 3*)	32
Section 3.2: 61 – 74 Squared (difficulty 3*)	35
Section 3.3: 26 - 39 Squared (difficulty 4*)	37
Section 3.4: 111 – 124 Squared (difficulty 3*)	40
Section 3.5: 76 – 89 Squared (difficulty 4*)	43
Section 3.6: Summary (difficulty 4*)	46
Chapter 4: The Squares of 101 to 999 (Optional Reading)	51
Section 4.1: 101 – 199 Squared (difficulty 4*)	51
Section 4.2: 201 – 999 Squared (difficulty 5*)	52
Chapter 5: Mathematical Proof of the Magic Square Quick Formula	55
Section 5.1: Proof of the Magic Square Quick Formula 1.1 with Ones Digit of 5 (difficulty 5*)	55
Section 5.2: Proof of the Magic Square Quick Formula 2.1 for Numbers around 50 (difficulty 5*)	55
Section 5.3: Proof of the Magic Square Quick Formula 2.2 for Numbers around 100 (difficulty 5*)	56
Section 5.4: Proof of the Magic Square Quick Formula 3.1 for 11 – 24 (difficulty 5*)	56
Section 5.5: Proof of Magic Square Quick Formula 4.1 for 100 – 1000 (difficulty 5*)	57

Part II: Magic Multiplication Speed Algorithm — 59
Chapter 6: Magic Multiplication of Any Two Numbers (Universal Formula) — 61
    Section 6.1: Universal Formula for Magic Multiplication (difficulty 4*) — 61
    Section 6.2: Magic Multiplication of Two Numbers When They Differ by < 20 (difficulty 4*) — 66
    Section 6.3: Magic Multiplication of Two Numbers with Both Odd or Even Tens Digit and the Same Number for the Ones Digit — 67

Chapter 7: Multiplication of Special Numbers (Special Formulas) — 69
    Section 7.1: Multiplication of Two Numbers Where the Tens Digit is the Same and the Sum of the Ones Digits Equals 10 (difficulty 2*) — 69
    Section 7.2: Multiplication of Two Numbers Where the Tens Digit Differs by 1 and the Sum of the Ones Digits Equals 10 (difficulty 3*) — 72
    Section 7.3: Multiplication of Two Numbers Where Sum of the Tens Digits Equals 10 and the Ones Digits are the same (difficulty 2*) — 75
    Section 7.4: Multiplication of Any Two Numbers Between 11 and 19 (difficulty 4*) — 78
    Section 7.5: Multiplication of Any Two Numbers Between 101 and 109 (difficulty 3*) — 81
    Section 7.6: Multiplication of Any Two Numbers Between 99 and 91 (difficulty 4*) — 84
    Section 7.7: Multiplication of Any Two Numbers Between 81 and 119 (optional) (difficulty 5*) — 87

Chapter 8: Proof of the Quick Formula for Magic Multiplication (difficulty 5*) — 89
    Section 8.1: Proof of Formula 6.1 for the Universal Multiplication — 89
    Section 8.2: Proof of Formula 7.1 for the Multiplication of Two Numbers Where the Tens Digits are the Same and the Sum of the Ones Digits Equals 10 — 89
    Section 8.3: Proof of Formula 7.2 for the Multiplication of Two Numbers Where the Tens Digit Differs by 1 and the Sum of the Ones Digits Equals 10 — 90
    Section 8.4: Proof of Formula 7.3 for the Multiplication of Two Numbers Where the Sum of the Tens Digits Equals 10 and the Ones Digits are the same — 90

Section 8.5: Proof of Formula 7.4 for the Multiplication of
    Any Two Numbers between 11 and 19   91
Section 8.6: Proof of Formula 7.5 for the Multiplication of
    Two Numbers Around 100   91

Part III: Magic Square Root Speed Algorithm   93
Chapter 9: Square Roots Within 1000000 (difficulty 5*)   95
Section 9.1: Ideas and Principles for Calculating Approximate
    Square Roots (difficulty 5*)   96
Section 9.2: Square Roots from 100 to 10000 (difficulty 5*)   99
Section 9.3: Square Roots from 1 to 100 (difficulty 5*)   101
Section 9.4: Square Roots from 10000 to 1000000 (difficulty 5*)   103

Conclusion   105
Self-Assessment Questions   107
Concluding Remarks   108
About the Author   109

HOW TO USE THIS BOOK

The two core elements of this book are:

1. The complexity of the calculation of squares can be greatly reduced by converting the calculation of the square of any two-digit number into the calculation of the sum of squares and addition and subtraction involving the number. This can be done through two simple calculation formulas. Anyone with a middle school math foundation can do this in their head.
2. You can transform any two digit multiplication into the calculation of squaring and the calculation of addition and subtraction with a simple quick formula. This quick calculation formula is the essence of the book.

As long as you are proficient in these three simple formulas, you can mentally calculate the square of any two-digit number and the product of any two-digit number.

Also included is a quick calculation method to approximate square roots, which is simple, easy to learn, and has a low threshold. As long as you have the basics of elementary school arithmetic, you can learn it. I haven't seen a rapid calculation that approximates square roots in any other books, and this method may be a breakthrough for rapid calculations.

The characteristic of rapid calculation is a compromise between the scope of application and computational efficiency. For numbers with certain characteristics, dedicated formulas are more computationally efficient, but have a narrower scope. The general formula is less efficient but has a wide range of applications. This book introduces both specialized formulas and general formulas. In general, we prefer to use specialized formulas. When no special formula is available, the general formula is used.

This book is divided into four parts.

Part I introduces the rapid calculation for squaring. It first introduces a popular method of square rapid calculation in Chapter 1 that is only applicable to a specific number of square rapid calculations in order to arouse your interest. The general calculation of squaring is divided into three chapters (Chapter 2 to Chapter 4) according to the degree of complexity. Chapter 5 is the mathematical proof of the formulas in this part.

Chapter 2 deals with the algorithm for squaring simple numbers, or squaring without carrying.

Chapter 3 deals with squaring with carrying.

Chapter 4 is the square rapid calculation for three-digit numbers. This rapid calculation formula is also the original creation of the author of this book.

Chapter 5 is the mathematical proof of all the square rapid calculation formulas, perfect for those who have a mathematical foundation in junior high school or above.

Part II introduces two-digit rapid multiplication, divided into three chapters (Chapters 6 to 8).

Chapter 6 introduces the general formulas for rapid multiplication. It applies to the multiplication of any two numbers and is one of the core contributions of this book.

Chapter 7 introduces special formulas for rapid multiplication and gives many examples to demonstrate the origin and usage of rapid multiplication formulas.

Chapter 8 gives mathematical proofs for all formulas of rapid multiplication.

Part III is the rapid calculation of approximate square roots and is the original creation of the author of this book.

Part IV is brief summary of the book. We can use a capital Y shape to describe the essence of this book.

The lower half of the Y is the branch, which is composed of the square rapid calculation of Part I.

The upper left of the Y is the multiplication algorithm (see Part II), which is supported by the lower part of the Y. We need to convert multiplication to squares in order to calculate the answer.

The upper right of the Y is the square root rapid calculation (see Part III), which is supported by the lower part of Y. Square roots are the inverse operation of squares. Use the square approximation method to find an approximation to the square root.

For the convenience of readers with different mathematical backgrounds, this book defines the difficulty degrees of each section, denoted by stars "*". The more "*", the more difficult it is. I suggest you select the "*" level according to your own mathematical level, and then select the corresponding chapter to read, so as to ensure you get the most out of this book.

SELF-ASSESSMENT QUESTIONS

Do you accept the challenge? Below are 60 self-assessment questions. Without using a calculator and without using pen and paper see how many you can answer. Record the time and the number of questions answered correctly.

65 x 65 =	85 x 85 =	58 x 58 =	53 x 53 =
47 x 47 =	45 x 45 =	108 x 108 =	103 x 103 =
98 x 98 =	92 x 92 =	12 x 12 =	18 x 18 =
22 x 22 =	38 x 38 =	28 x 28 =	68 x 68 =
89 x 89 =	72 x 72 =	82 x 82 =	112 x 112 =
103 x 108 =	101 x 107 =	13 x 18 =	14 x 16 =
96 x 99 =	92 x 97 =	53 x 57 =	66 x 64 =
24 x 84 =	36 x 76 =	33 x 73 =	49 x 69 =
18 x 98 =	44 x 64 =	46 x 54 =	57 x 63 =
28 x 32 =	78 x 82 =	37 x 43 =	61 x 79 =
58 x 62 =	54 x 66 =	51 x 67 =	49 x 51 =
39 x 43 =	37 x 45 =	58 x 98 =	21 x 81 =
19 x 89 =	34 x 94 =	25 x 105 =	32 x 72 =
152 x 152 =	198 x 198 =	302 x 302 =	499 x 499 =
$\sqrt{92}$ =?	$\sqrt{35}$ =?	$\sqrt{8088}$ =?	$\sqrt{3028}$ =?

Name: _____ Time: _____ minutes
Number of Questions Answered Correctly: _____
Date: Year ___ Month ___ Day ___

If you can answer 20 questions correctly in 30 minutes, that is very good. But when you finish this book, you should be able to answer more than 50 questions correctly in only 30 minutes.

PART I
MAGIC SQUARE SPEED ALGORITHM

The core of the square rapid calculation is to convert the calculation of the square of a larger number into two calculations: the calculation of the square of a smaller number and addition and subtraction.

Generally speaking, the simpler the rapid calculation formula of the quick algorithm, the smaller its scope of application. Formula 1.1 as introduced in Section 1.2 applies only to nine numbers less than 100. Can we find a rapid calculation formula that both has a wider scope and is concise?

After reading the first three chapters of Part I, you will be able to mentally square all numbers up to 125 using two simple formulas. You can master them in just a few hours.

This part is especially suitable for students in the upper grades of elementary school or middle school students. when you understand the principles behind rapid calculation, it will increase your interest and confidence in learning mathematics, and in your future study, it can save you a lot of time. This section also applies to parents.

By the end of Chapter 4, you can mentally square all numbers up to 1,000. This part is a bit more difficult and is meant for optional reading.

Once you have done Chapter 5, "you will understand the mathematics behind the quick formula. This chapter is for those with junior high school math or above.

CHAPTER 1: INTRODUCTION

Section 1.1: Background Knowledge – Nine-by-Nine Multiplication Table (Difficulty 1*)

1 x 1 = 1								
1 x 2 = 2	2 x 2 = 4							
1 x 3 = 3	2 x 3 = 6	3 x 3 = 9						
1 x 4 = 4	2 x 4 = 8	3 x 4 = 12	4 x 4 = 16					
1 x 5 = 5	2 x 5 = 10	3 x 5 = 15	4 x 5 = 20	5 x 5 = 25				
1 x 6 = 6	2 x 6 = 12	3 x 6 = 18	4 x 6 = 24	5 x 6 = 30	6 x 6 = 36			
1 x 7 = 7	2 x 7 = 14	3 x 7 = 21	4 x 7 = 28	5 x 7 = 35	6 x 7 = 42	7 x 7 = 49		
1 x 8 = 8	2 x 8 = 16	3 x 8 = 24	4 x 8 = 32	5 x 8 = 40	6 x 8 = 48	7 x 8 = 56	8 x 8 = 64	
1 x 9 = 9	2 x 9 = 18	3 x 9 = 27	4 x 9 = 36	5 x 9 = 45	6 x 9 = 54	7 x 9 = 63	8 x 9 = 72	9 x 9 = 81

The 9 x 9 multiplication table is the foundation of the foundation. It is here as a warm-up.

Section 1.2: Fun Facts – Squares of Numbers Whose Ones Digit is 5 (difficulty 2*)

Basics: The reader is assumed to be able to do simple addition, use the nine-by-nine multiplication table, and understand the concepts of multiplication and squaring.

Before reading this chapter, try to quickly calculate the following results. If you can do it quickly, skip this section and go straight to Chapter 2.

$15 \times 15 = 15^2 =$ $65 \times 65 = 65^2 =$ $35 \times 35 = 35^2 =$

$95 \times 95 = 95^2 =$ $55 \times 55 = 55^2 =$ $25 \times 25 = 25^2 =$

$45 \times 45 = 45^2 =$ $75 \times 75 = 75^2 =$ $85 \times 85 = 85^2 =$

If your calculation is not fast enough, read on.

What is the square of a number? It is the number multiplied by itself. For example, 45 squared is 45 x 45, expressed as 45^2.

When I was in junior high school, my math teacher mysteriously asked us, can you quickly calculate the square of 45 mentally? If you can, can you quickly square any number with ones digit of 5 mentally? Our classmates and I looked at each other in dismay, and no one answered. Let's look at the mystery of the calculation of squares.

We usually use the following vertical calculations:

```
      4 5
×     4 5
    ─────
      2 2 5
    1 8 0
    ─────
    2 0 2 5
```

It takes 4 steps of one-digit multiplication and 4 steps of addition (a total of 8 steps) to calculate the square, which is difficult to do with mental arithmetic. But my math teacher taught us a trick: square it in just two steps. Let me show you how.

See these examples:
$15 \times 15 = 15^2 = 225$
$25 \times 25 = 25^2 = 625$
$35 \times 35 = 35^2 = 1225$
$45 \times 45 = 45^2 = 2025$
$55 \times 55 = 55^2 = 3025$
$65 \times 65 = 65^2 = 4225$
$75 \times 75 = 75^2 = 5625$
$85 \times 85 = 85^2 = 7225$
$95 \times 95 = 95^2 = 9025$

Can you see the pattern? Let's rewrite these examples in a different way:
15 × **15** = **15**^{2} = (**1** ×(**1** + 1)) × 100 + 25 = (**1** × **2**) × 100 + 25 = 2 25
25 × **25** = **25**^{2} = (**2** ×(**2** + 1)) × 100 + 25 = (**2** × **3**) × 100 + 25 = 6 25
35 × **35** = **35**^{2} = (**3** ×(**3** + 1)) × 100 + 25 = (**3** × **4**) × 100 + 25 = 12 25
45 × **45** = **45**^{2} = (**4** ×(**4** + 1)) × 100 + 25 = (**4** × **5**) × 100 + 25 = 20 25
55 × **55** = **55**^{2} = (**5** ×(**5** + 1)) × 100 + 25 = (**5** × **6**) × 100 + 25 = 30 25
65 × **65** = **65**^{2} = (**6** ×(**6** + 1)) × 100 + 25 = (**6** × **7**) × 100 + 25 = 42 25
75 × **75** = **75**^{2} = (**7** ×(**7** + 1)) × 100 + 25 = (**7** × **8**) × 100 + 25 = 56 25
85 × **85** = **85**^{2} = (**8** ×(**8** + 1)) × 100 + 25 = (**8** × **9**) × 100 + 25 = 72 25
95 × **95** = **95**^{2} = (**9** ×(**9** + 1)) × 100 + 25 = (**9** × **10**) × 100 + 25 = 90 25

Now we can see the pattern more clearly! We can describe this law as follows:

Because the ones digit of this number is 5 (a constant), we only need to remember the tens digit of the number.

The rule for calculating the square of this number is:

The first two digits are the tens digit of the number multiplied by the number one larger than the tens digit. The last two digits are always 25.

The semi-formal definition of this rapid calculation is as follows: let us call the tens-digit (changing number) the "magic number". The ones digit is always 5. Thus, the square of a number can be calculated like this:

Formula 1.1: (some number)^2 = magic number x (magic number + 1) x 100 + 25

Formula 1.1 can be explained as follows: if the ones digit of a number is 5, we can call the tens digit a "magic number." The first two digits of the square of the number is equal to the magic number multiplied by a number that is 1 larger than it. The last two digits are always 25 (constant). So, when we calculate 45^2 mentally, we just need to remember that the magic number is 4, and the number 1 greater than 4 is 5. We know that the first two digits of the square are 4 x 5 = 20, and the last two digits are always 25. Thus, 45^2 = 2025. Similarly, we can quickly calculate 65^2 = (6 x 7) x 100 + 25 = 4225.

This method reduces the traditional eight steps to two steps. It is simple and the calculation speed is as fast as magic. So I call this type of square rapid calculation, "magic square." Extending the rapid squaring algorithm to rapid multiplication, I call it "magic multiplication." Hence the title of this book.

For the convenience of our discussion later, we formalize Formula 1.1 as follows.

We call a number A and the magic number m, where A = 10 * m + 5,

Formula 1.1: A^2 = (10 * m + 5)^2 = m * (m + 1) x 100 + 25

It should be noted that in this book, when two numbers are multiplied, the multiplication symbol is represented by "x". When multiplying two letters (or a number and a letter), the multiplication symbol is represented by "*".

To see why Formula 1.1 is correct, see Chapter 5 (Section 5.1) for the mathematical proof of the magic square quick formula.

Now, you can quickly mentally work out

35 x 35 = 35^2 = (3 x (3 + 1)) x 100 + 25 = (3 x 4) x 100 + 25 = 1225

And

75 x 75 = 75^2 = (7 x (7 + 1)) x 100 + 25 = (7 x 8) x 100 + 25 = 5625.

In this formula, we find the magic number (tens digit) in the first step, the magic number + 1 (the tens digit plus 1) in the second step, and the magic number * (magic number + 1) in the third step. The first two digits of the calculation result are the magic number multiplied by the (magic number + 1), and the last two digits are always 25, so no calculation is required.

According to the calculation steps, the "find a number" in the first step is not a calculation, the "add 1" in the second step is a half-step calculation, and the third step is a "calculation step". This algorithm only uses one step of calculation, at most 1.5 steps of calculation, which simplifies the traditional 8-step calculation to 1.5 steps.

We abstract Formula 1.1 into three steps of calculation:
Step 1: Find the magic number (tens);
Step 2: Calculate the magic number* (magic number + 1) and put it in the first two digits (if there is only one digit, put a zero in the front as a placeholder);
Step 3: Always put 25 in the last two digits.

The core of rapid calculation is to simplify the algorithm as much as possible. I hope readers will find more mysteries in the follow-up reading.

This chapter is just a warm-up, and the core content will start with the next chapter.

Exercise 1: In the following exercises, please identify the "magic number" of each question, and then calculate the answer.

$25^2 =$ $55^2 =$

$45^2 =$ $75^2 =$

$95^2 =$ $65^2 =$

$85^2 =$ $15^2 =$

$35^2 =$

CHAPTER 2: SQUARE NUMBERS WITHOUT CARRYING

Assumed knowledge: The reader is assumed to know how to use negative numbers.

We learned the rapid calculation method of the square of a number when the ones digit is 5. I asked my school math teacher if there was a rapid calculation method for the square of a number when the tens digit was 5? He told me that he didn't know. Later, I not only found a quick way to calculate the square of numbers whose tens digit is 5, but also found a quick way to calculate the square of any two-digit number. Still later (in 2011), I found a quick way to multiply any two-digit number (see Part II). In 2017, a rapid calculation method for calculating the square for a number within 1000 was found (see Chapter 4). And in 2021, a rapid calculation method for calculating the square root of less than 1,000,000 was found (see Part III). The methods presented in this book have taken me 50 years to compile.

Before explaining this chapter, please try to quickly calculate the following results. If you can, skip this chapter and go straight to Chapter 3.

$55^2 =$	$45^2 =$	$105^2 =$	$95^2 =$
$56^2 =$	$46^2 =$	$106^2 =$	$96^2 =$
$58^2 =$	$48^2 =$	$108^2 =$	$98^2 =$
$54^2 =$	$44^2 =$	$104^2 =$	$94^2 =$
$52^2 =$	$42^2 =$	$102^2 =$	$92^2 =$
$51^2 =$	$41^2 =$	$101^2 =$	$91^2 =$
$57^2 =$	$47^2 =$	$107^2 =$	$97^2 =$
$53^2 =$	$43^2 =$	$103^2 =$	$93^2 =$
$59^2 =$	$49^2 =$	$109^2 =$	$99^2 =$

If your calculations were not fast enough, read on.

This chapter explains how to square some special numbers, such as $54^2 = 2916$, $46^2 = 2116$, $104^2 = 10816$, $96^2 = 9216$. From now on, we will express 54 x 54 as 54^2, 46 x 46 as 46^2, and so on.

This chapter is divided into 5 sections, introducing two algorithms respectively.

Section 2.1: 51 - 59 Squared (difficulty 2*)

Please see the examples below:

$51^2 = 26\ 01$
$52^2 = 27\ 04$
$53^2 = 28\ 09$
$54^2 = 29\ 16$
$55^2 = 30\ 25$
$56^2 = 31\ 36$
$57^2 = 32\ 49$
$58^2 = 33\ 64$
$59^2 = 34\ 81$

Let us convert it a little bit:

$51^2 = 26\ 01 = (25 + 1) \times 100 + 1^2$
$52^2 = 27\ 04 = (25 + 2) \times 100 + 2^2$
$53^2 = 28\ 09 = (25 + 3) \times 100 + 3^2$
$54^2 = 29\ 16 = (25 + 4) \times 100 + 4^2$
$55^2 = 30\ 25 = (25 + 5) \times 100 + 5^2$
$56^2 = 31\ 36 = (25 + 6) \times 100 + 6^2$
$57^2 = 32\ 49 = (25 + 7) \times 100 + 7^2$
$58^2 = 33\ 64 = (25 + 8) \times 100 + 8^2$
$59^2 = 34\ 81 = (25 + 9) \times 100 + 9^2$

To summarize, the first two digits of the result are 25 plus the ones digit, and the last two digits of the result are the square of the ones digit. For example, $56^2 = (25 + 6) \times 100 + 6^2 = 31\ 36$

Because the tens digits of the numbers above are all 5 (constant), the ones digit (variable number) is the magic number. As long as we remember this magic number, we can easily calculate the square. That is to say, the first two digits are 25 (constant) plus the magic number, and the last two digits are the square of the magic number.

Note that the square of the magic number must always occupy the last two digits. We should put a zero in front of the square as a placeholder if the square is a single digit. For example, 52 = 50 + 2, so the magic number of 52 is 2. $52^2 = (25 + 2) \times 100 + 2^2 = 27\ 04$, not 274.

Any number between 51 - 59 can be expressed as (50 + magic number), so, (some number)² can be expressed as follows:

Formula 2.1: (some number)² = (50 + ones digit)² = (25 + ones digit) x 100 + (ones digit)² or

Formula 2.1: (some number)² = (50 + magic number)² = (25 + magic number) x 100 + (magic number)²

Note that (25 + magic number) x 100 means that (25 + magic number) takes the first two digits. This calculation also requires only one step of addition and one step of multiplication, and mental arithmetic is naturally fast.

Please note that the magic number changes with the formula. In Chapter 2, the magic number is the tens digit of the number, and in this section, the magic number is the ones digit of the number. The basic principle is that the magic number represents the numbers that change. Don't get confused!

We formally express Formula 2.1 as follows:

We call the number A and the magic number m, where A = 50 + m, so m = A − 50,

Formula 2.1: $A^2 = (50 + m)^2 = (25 + m) \times 100 + m^2$

We derived Formula 2.1 from experience, and while it may not seem intuitive at all, it is absolutely correct. Interested readers can refer to Chapter 5 (Section 5.2), which has the mathematical proof of the magic number square quick formula.

In fact, in this formula, we find the magic number (ones digit) in the first step, 25 + the magic number in the second step, and the magic number squared in the third step. The first two digits of the calculation result are 25 + the magic number, and the last two digits are always the square of the magic number. Because the first step to "find a number" is not a calculation step, it is said that Formula 2.1 only uses two simple calculations (an addition and a one-digit square). Thus, the traditional 8-step calculation of two-digit multiplication is reduced to a 2-step calculation.

When a certain number is between 51 and 59, we can abstract Formula 2.1 into the following three steps:

Step 1: Find the magic number (ones digit), the magic number is between 1 - 9;
Step 2: Calculate (25 + magic number) and put it in the first two digits;
Step 3: Calculate the square of the magic number and put it in the last two digits (if there is only one digit, use 0 in the front).

You may have noticed that there are two rapid calculations for 55^2: one is to use the Formula 1.1: $55^2 = (5 \times (5 + 1)) \times 100 + 25 = 30\ 25$ (see Section 1.2), which takes 1.5 steps to calculate; another method is to use Formula 2.1: $55^2 = (25 + 5) \times 100 + 5^2 = 30\ 25$, which takes 2 steps to calculate. Both rapid calculation methods are simple, and you can use either method.

Exercise 2.1: In the following exercises, please work out the magic number in each question, and then calculate the answer.

$52^2 =$ $56^2 =$ $58^2 =$ $51^2 =$

$59^2 =$ $57^2 =$ $54^2 =$ $53^2 =$

Section 2.2: 41 - 49 Squared (difficulty 3*)

Assumed knowledge: This chapter assumes the reader knows how to work with negative numbers.

Please see the examples below:

$49^2 = 24\ 01$
$48^2 = 23\ 04$
$47^2 = 22\ 09$
$46^2 = 21\ 16$
$45^2 = 20\ 25$
$44^2 = 19\ 36$
$43^2 = 18\ 49$
$42^2 = 17\ 64$
$41^2 = 16\ 81$

From the above examples, the pattern is not obvious. Let's change the above examples a little bit:

$49^2 = (50 - 1)^2 = (50 + (-1))^2 = 24\ 01 = (25 + (-1)) \times 100 + (-1)^2$
$48^2 = (50 - 2)^2 = (50 + (-2))^2 = 23\ 04 = (25 + (-2)) \times 100 + (-2)^2$
$47^2 = (50 - 3)^2 = (50 + (-3))^2 = 22\ 09 = (25 + (-3)) \times 100 + (-3)^2$
$46^2 = (50 - 4)^2 = (50 + (-4))^2 = 21\ 16 = (25 + (-4)) \times 100 + (-4)^2$
$45^2 = (50 - 5)^2 = (50 + (-5))^2 = 20\ 25 = (25 + (-5)) \times 100 + (-5)^2$
$44^2 = (50 - 6)^2 = (50 + (-6))^2 = 19\ 36 = (25 + (-6)) \times 100 + (-6)^2$
$43^2 = (50 - 7)^2 = (50 + (-7))^2 = 18\ 49 = (25 + (-7)) \times 100 + (-7)^2$
$42^2 = (50 - 8)^2 = (50 + (-8))^2 = 17\ 64 = (25 + (-8)) \times 100 + (-8)^2$
$41^2 = (50 - 9)^2 = (50 + (-9))^2 = 16\ 81 = (25 + (-9)) \times 100 + (-9)^2$

Think about it, what is the magic number this time?

Yes, the magic number is the difference between the number and 50. For example, the magic number of 48 is 48 - 50 = -2, not the ones digit 8. Here, novice readers are especially prone to confusion. You are advised to pay special attention here, or the results will be wrong. Also, please remember that the square of any negative number is the square of the same positive number. For example, $(-6)^2 = 6^2 = 36$.

Likewise, as long as you remember this magic number, you can easily calculate the desired square. That is to say, the first two digits are 25 plus the magic number, and the last two digits are the square of the magic number. For example, the magic number for 48 is 48 – 50 = -2. So, $48^2 = (25 + (-2)) \times 100 + (-2)^2 = 23\ 04$.

On a closer look, did you notice that the approach in this section is the same as the approach in section 2.1? Check out the following two examples:

Described in Section 2.1: $58^2 = (50 + \mathbf{8})^2 = (25 + \mathbf{8}) \times 100 + \mathbf{8}^2 = 33\ 64$

This section introduces: $42^2 = (50 + (\mathbf{-8}))^2 = (25 + (\mathbf{-8})) \times 100 + (\mathbf{-8})^2 = 17\ 64$

The attentive reader may have noticed that in the two examples, the magic number of 58 in the first example is 8, and the magic number of 42 in the second example is -8. The formula for calculating squares are the same. You should pay special attention here. Only remember the magic number, forget the ones digit. For example, the magic number of 42 is -8, so there is no need to think that the ones digit is 2, otherwise it is easy to make mistakes.

In fact, the method introduced in Section 2.1 and this section is the same. That is, the magic number is the difference between this number and 50. The only difference is: magic numbers in section 2.1 are positive (e.g. 57 - 50 = 7), magic numbers in this section are negative (e.g. 43 - 50 = -7). The method for calculating the square of numbers from 51 - 59 is exactly the same as the method for calculating the square of numbers from 49 - 41. The only difference is that the magic number is positive versus negative.

We calculate 43^2 using Formula 2.1: A = 43, m = 43 – 50 = -7, then we have

$$43^2 = (50 + (-7))^2 = (25 + (-7)) \times 100 + (-7)^2 = 18\ 49$$

In this formula, we calculate the magic number in the first step (use the number – 50). In the second step, we calculate 25 + the magic number. In the third step, we calculate the magic number squared. The first two digits of the calculation are (25 + magic number), and the last two digits are always the square of the magic number. So, Formula 2.1 uses 3 simple steps here (instead of the two steps in the previous section), simplifying the traditional 8-step calculation of two-digit multiplication to a 3-step calculation.

When a certain number is between 49 - 41, we abstract Formula 2.1 into the following three steps:

Step 1: Calculate the magic number (a certain number - 50), the magic number is between (-1) - (-9):
Step 2: Calculate (25 + magic number) and put it in the first two digits;
Step 3: Calculate the square of the magic number and put it in the last two digits (if there is only one digit, use 0 in the front).

Compared to Section 2.1, the first step is to calculate the magic number, not to find the magic number.

Note that there are two formulas for 45^2. The method of Formula 1.1, introduced in Section 1.2 is: $45^2 = (4 \times (4 + 1)) \times 100 + 25 = 20\ 25$, the magic number is 4. It is calculated in 1.5 steps and has no negative numbers. The method of Formula 2.1, introduced in this section (2.2) is $45^2 = (25 + (-5)) \times 100 + (-5)^2 = 20\ 25$, the magic number is -5. It took 3 steps, and there was a negative number operation. You can use either method, but Formula 1.1 is slightly simpler.

Exercise 2.2: In the following exercises, please first calculate the magic number of each question, and then calculate the answer.

Please note: it is especially important to calculate the magic number first in this section. For example, the magic number for 42 is -8 instead of 2.

$42^2 =$ $44^2 =$ $41^2 =$

$47^2 =$ $49^2 =$ $46^2 =$

$45^2 =$ $48^2 =$ $43^2 =$

Section 2.3: 101 - 109 Squared (difficulty 2*)

Please see the examples below:

$101^2 = 102\ 01$
$102^2 = 104\ 04$
$103^2 = 106\ 09$
$104^2 = 108\ 16$
$105^2 = 110\ 25$
$106^2 = 112\ 36$
$107^2 = 114\ 49$
$108^2 = 116\ 64$
$109^2 = 118\ 81$

Do you see the pattern? What is the magic number? It will be clear after reading the following.

$101^2 = 102\ 01 = (101 + \mathbf{1}) \times 100 + \mathbf{1}^2$
$102^2 = 104\ 04 = (102 + \mathbf{2}) \times 100 + \mathbf{2}^2$
$103^2 = 106\ 09 = (103 + \mathbf{3}) \times 100 + \mathbf{3}^2$
$104^2 = 108\ 16 = (104 + \mathbf{4}) \times 100 + \mathbf{4}^2$
$105^2 = 110\ 25 = (105 + \mathbf{5}) \times 100 + \mathbf{5}^2$
$106^2 = 112\ 36 = (106 + \mathbf{6}) \times 100 + \mathbf{6}^2$
$107^2 = 114\ 49 = (107 + \mathbf{7}) \times 100 + \mathbf{7}^2$
$108^2 = 116\ 64 = (108 + \mathbf{8}) \times 100 + \mathbf{8}^2$
$109^2 = 118\ 81 = (109 + \mathbf{9}) \times 100 + \mathbf{9}^2$

In the above calculations, the result of the square is 5 digits. The magic number is the difference between the number itself and 100. For example, the magic number of 106 is 106 – 100 = 6, which is the ones digit of the number. Remember this magic number, and it's simple to calculate the square: the first three digits are the number itself plus the ones digit (magic number), and the last two digits are the square of the ones digit (magic number). Be careful not to choose the wrong digits! For example: $101^2 = 10201$ instead of 121.

Formula 2.2: (the number)^2 = (100 + ones digit)^2 = (the number + ones digit) x 100 + (ones digit)^2

or

Formula 2.2: (the number)^2 = (100 + magic number)^2 = (the number + magic number) x 100 + (magic number)^2

We formally express Formula 2.2 as follows. We call the number A and the magic number m, where A = 100 + m, so m = A − 100,

Formula 2.2: $A^2 = (100 + m)^2 = (A + m) \times 100 + m^2$

We calculate 108^2 using Formula 2.2 as follows: A = 108, m = 108 − 100 = 8 (ones digit). Applying Formula 2.2 we get:

$$108^2 = (108 + 8) \times 100 + 8^2 = 116\ 64$$

Now we know how to quickly calculate 106^2 mentally. The magic number of 106 is 106 − 100 = 6. So, the first three digits of 106^2 are 106 + 6 = 112, and the last two digits are (magic number)^2, that is, $6^2 = 36$. So, $106^2 = 112\ 36$.

For a detailed proof, see Chapter 5 (Section 5.3), Mathematical Proof of the Magic Square Quick Formula.

In this formula, the first step is to find the magic number (ones digit), the second step is to calculate (the number + the magic number), and the third step is to calculate the magic number squared. The first three digits of the calculation result are (the number + the magic number), and the last two digits are always the square of the magic number. Because finding the ones digit in the first step is not a calculation step, here we simplify the 14-step (9-step multiplication and 5-step addition) calculation of the traditional three-digit multiplication to a 2-step calculation.

When a certain number is between 101 - 109, we abstract Formula 2.2 into the following three steps:

Step 1: Find the magic number (ones digit), the magic number is between 1 - 9;
Step 2: Calculate (the number + magic number) and put it in the first three digits;
Step 3: Calculate the square of the magic number and place it in the last two digits (if there is only one digit, use 0 in the front).

105^2 has two algorithms. The Formula 1.1 introduced in Chapter 1 is: $105^2 = (10 \times (10 + 1)) \times 100 + 25 = 11025$, the magic number is 10, and it contains 1.5 steps of calculation. The Formula 2.2 introduced in this section (2.3) is: $105^2 = (105 + 5) \times 100 + 5^2 = 11025$, the magic number is 5, which is a 2-step calculation. Both algorithms are easy.

Exercise 2.3: In the following exercises, please work out the magic number for each question, and then calculate the answer.

$107^2 =$ $102^2 =$ $103^2 =$

$108^2 =$ $105^2 =$ $106^2 =$

$101^2 =$ $109^2 =$ $104^2 =$

Section 2.4: 91 - 99 Squared (difficulty 3*)

Assumed knowledge: The reader is assumed to know how to operate on negative numbers.

Please see the examples below:

$99^2 = 98\ 01$
$98^2 = 96\ 04$
$97^2 = 94\ 09$
$96^2 = 92\ 16$
$95^2 = 90\ 25$
$94^2 = 88\ 36$
$93^2 = 86\ 49$
$92^2 = 84\ 64$
$91^2 = 82\ 81$

Do you notice the pattern? Please read the following part:

$99^2 = 98\ 01 = (99 + (-1)) \times 100 + (-1)^2$
$98^2 = 96\ 04 = (98 + (-2)) \times 100 + (-2)^2$
$97^2 = 94\ 09 = (97 + (-3)) \times 100 + (-3)^2$
$96^2 = 92\ 16 = (96 + (-4)) \times 100 + (-4)^2$
$95^2 = 90\ 25 = (95 + (-5)) \times 100 + (-5)^2$
$94^2 = 88\ 36 = (94 + (-6)) \times 100 + (-6)^2$
$93^2 = 86\ 49 = (93 + (-7)) \times 100 + (-7)^2$
$92^2 = 84\ 64 = (92 + (-8)) \times 100 + (-8)^2$
$91^2 = 82\ 81 = (91 + (-9)) \times 100 + (-9)^2$

We define the magic number as the difference between the number and 100. For example, the magic number for 97 is 97 − 100 = -3, not the ones digit 7.

The pattern of squares is: after we calculate the magic number, the first two digits of the square of a number are the number itself plus the magic number, and the last two digits are the square of the magic number. As long as the magic number is memorized (forget the ones digit), the result is easy to calculate.

You may notice that section 2.3 is very similar to this section (2.4). The magic number is the difference between the number itself and 100. The only difference is that one is positive and the other is negative.

We know that the magic number of 96 is m = 96 − 100 = -4. So, the first two digits of 96^2 are 96 + (-4) = 92, and the last two digits are (magic number)^2, that is $(-4)^2$ = 16. So, 96^2 = 92 16.

In Formula 2.2, we calculate the magic number (the number - 100) in the first step, (the number + magic number) in the second step, and then calculate the square of the magic number in the third step. The first two digits of the calculation result are (the number + the magic number), and the last two digits are always the square of the magic number. Therefore, here we have simplified the traditional 8-step calculations of two-digit multiplication to 3 steps of calculations.

Please note two points:

(1) The magic number is not a ones digit and is a negative number;
(2) Because the magic number is negative, the calculation of (the number + magic number) is actually a subtraction.

When a certain number is between 99 - 91, we abstract **Formula 2.2** into the following three steps:

Step 1: Calculate the magic number (the number - 100). The magic number is between (-1) - (-9);
Step 2: Calculate (the number + the magic number) and put it in the first two digits;
Step 3: Calculate the square of the magic number and place it in the last two digits (if there is only one digit, use 0 in the front).

Unlike Section 2.3, the first step here is to calculate the magic number, not to work out the magic number.

Also, the 95^2 has two algorithms. The algorithm of Section 1.2 Formula 1.1 is: 95^2 = 9 x (9 + 1) x 100 + 25 = 90 25, and the magic number is 9. This algorithm consists of 1.5 simple steps, with no negative operations. The Formula 2.2 in this section is: 95^2 = (95 + (−5)) x 100 + $(-5)^2$ = 90 25, and the magic number is -5, including 3 steps of calculation, and the negative number operation. It is slightly simpler to use Formula 1.1.

So far, we have learned three formulas. The first step in each formula is to calculate (or find) the magic number, and then calculate the result based on the magic number. This is the core idea of this book.

Exercise 2.4: In the following exercises, please calculate the magic number of each question first, and then calculate the answer.

$95^2 =$ $98^2 =$ $91^2 =$

$92^2 =$ $94^2 =$ $97^2 =$

$96^2 =$ $99^2 =$ $93^2 =$

Section 2.5: Summary (difficulty 3*)

In this chapter, we introduced two simple formulas to calculate some simple squares. The core idea of the formula is to convert the calculation of the square of a larger number into the calculation of the square of a smaller number and addition and subtraction, thereby reducing the complexity of the calculation and improving the speed.

Now, let us compare Formula 2.1 with Formula 2.2 to see how they are similar and how they are different.

$$\text{Formula 2.1: } A^2 = (50 + m)^2 = (25 + m) \times 100 + m^2$$

Formula 2.1 is only applicable to calculate the square of a certain number A (40 < A < 60) around 50, where the magic number $m = A - 50$ (-10 < m < 10).

$$\text{Formula 2.2: } A^2 = (100 + m)^2 = (A + m) \times 100 + m^2$$

Formula 2.2 is only applicable to calculate the square of a certain number A (90 < A < 110) around 100, where the magic number $m = A - 100$ (-10 < m < 10).

The two formulas are similar in that the first two digits (or first three digits) are the number plus the magic number, the last two digits are (magic number)², and the magic number m satisfies -10 < m < 10. The only difference is that the first two digits of the square of a certain number A around 50 are 25 + the magic number, while the first two (or three) digits of the square of a certain number A around 100 are "the number A + the magic number". Both formulas are extremely simple and easy to calculate.

Through the above two formulas, we convert a complex two-digit (or three-digit) square calculation into a simple addition and subtraction and a square less than 10. Do you feel the magic of the magic number square?

Before studying this book, you should be able to easily mentally calculate the squares of 1 - 9, and 10, 20, 30, 40, 50, 60, 70, 80, 90, and 100; by applying Formula 1.1, you will easily calculate the squares of 15, 25, 35, 45, 55, 65, 75, 85, 95 and 105; by applying Formula 2.1, you will easily calculate the squares of 41 - 49 and 51 - 59; by Formula 2.2, you will easily calculate 91 - 99 squared and 101 - 109 squared. Therefore, you can easily mentally square 61 numbers so far.

Can you now write the square of the 61 numbers mentioned above using only mental arithmetic? If you are not proficient, write it down several times. The squares of these 61 numbers will be used frequently in later chapters. By the time you finish the next chapter, you'll be able to square everything from 1 to 125, and more. Isn't speed really amazing? If you want to know more secrets, keep reading.

Comprehensive exercises for this chapter:

$55^2 =$	$45^2 =$	$95^2 =$	$105^2 =$
$56^2 =$	$46^2 =$	$96^2 =$	$106^2 =$
$58^2 =$	$48^2 =$	$98^2 =$	$108^2 =$
$54^2 =$	$44^2 =$	$94^2 =$	$104^2 =$
$52^2 =$	$42^2 =$	$92^2 =$	$102^2 =$
$51^2 =$	$41^2 =$	$91^2 =$	$101^2 =$
$57^2 =$	$47^2 =$	$97^2 =$	$107^2 =$
$53^2 =$	$43^2 =$	$93^2 =$	$103^2 =$
$59^2 =$	$49^2 =$	$99^2 =$	$109^2 =$

CHAPTER 3: SQUARE NUMBERS WITH CARRYING

Assumed knowledge: It is assumed that the reader is proficient in the content of the first two chapters, can do addition with carrying and subtraction with borrowing, and can especially make use of Formula 1.1 (see **Chapter 1**), Formula 2.1 (see **Section 2.1**) and Formula 2.2 (see **Section 2.3**).

Chapter 2 introduces two formulas that can be used to square simple numbers (without carrying), including 51 - 59 squared, 49 - 41 squared, 101 - 109 squared, and 99 - 91 squared. This chapter discusses how to use the same formula to square more numbers.

Before explaining this chapter, please try to quickly calculate the following. If you can do it, skip this chapter and go straight to Chapter 4.

$15^2 =$	$65^2 =$	$35^2 =$	$115^2 =$	$85^2 =$
$16^2 =$	$61^2 =$	$39^2 =$	$111^2 =$	$89^2 =$
$18^2 =$	$62^2 =$	$38^2 =$	$112^2 =$	$88^2 =$
$14^2 =$	$63^2 =$	$37^2 =$	$113^2 =$	$87^2 =$
$12^2 =$	$64^2 =$	$36^2 =$	$114^2 =$	$86^2 =$
$11^2 =$	$66^2 =$	$34^2 =$	$116^2 =$	$84^2 =$
$17^2 =$	$67^2 =$	$33^2 =$	$117^2 =$	$83^2 =$
$13^2 =$	$68^2 =$	$32^2 =$	$118^2 =$	$82^2 =$
$19^2 =$	$69^2 =$	$31^2 =$	$119^2 =$	$81^2 =$
$21^2 =$	$71^2 =$	$29^2 =$	$121^2 =$	$79^2 =$
$22^2 =$	$72^2 =$	$28^2 =$	$122^2 =$	$78^2 =$
$23^2 =$	$73^2 =$	$27^2 =$	$123^2 =$	$77^2 =$
$24^2 =$	$74^2 =$	$26^2 =$	$124^2 =$	$76^2 =$

If your calculations were not fast enough, read on. After reading this chapter, you should be able to do these calculations quickly.

Let's look at some examples first:

$19^2 = 361$, $69^2 = 4761$, $31^2 = 961$, $119^2 = 14161$, $81^2 = 6561$.

And more examples:

$13^2 = 169$, $63^2 = 3969$, $37^2 = 1369$, $113^2 = 12769$, $87^2 = 7569$.

These examples may seem difficult, but after you read this chapter, you will find it easy.

This chapter is divided into 5 sections. We will only introduce a new formula (**Formula 3.1**) and its applications.

Section 3.1: 11 - 24 Squared (difficulty 3*)

Please see the examples below:

$11^2 = 121$
$12^2 = 144$
$13^2 = 169$
$14^2 = 196$
$15^2 = 225$
$16^2 = 256$
$17^2 = 289$
$18^2 = 324$
$19^2 = 361$

Do you see the pattern? Now take a look at the following part:

If the above calculations are disassembled into the following form, we can see the pattern.

$11^2 = 121 = (11 + \mathbf{1}) \times 10 + \mathbf{1}^2 = 12 \times 10 + 1 = 120 + 1$
$12^2 = 144 = (12 + \mathbf{2}) \times 10 + \mathbf{2}^2 = 14 \times 10 + 4 = 140 + 4$
$13^2 = 169 = (13 + \mathbf{3}) \times 10 + \mathbf{3}^2 = 16 \times 10 + 9 = 160 + 9$
$14^2 = 196 = (14 + \mathbf{4}) \times 10 + \mathbf{4}^2 = 18 \times 10 + 16 = 180 + 16$
$15^2 = 225 = (15 + \mathbf{5}) \times 10 + \mathbf{5}^2 = 20 \times 10 + 25 = 200 + 25$
$16^2 = 256 = (16 + \mathbf{6}) \times 10 + \mathbf{6}^2 = 22 \times 10 + 36 = 220 + 36$
$17^2 = 289 = (17 + \mathbf{7}) \times 10 + \mathbf{7}^2 = 24 \times 10 + 49 = 240 + 49$
$18^2 = 324 = (18 + \mathbf{8}) \times 10 + \mathbf{8}^2 = 26 \times 10 + 64 = 260 + 64$
$19^2 = 361 = (19 + \mathbf{9}) \times 10 + \mathbf{9}^2 = 28 \times 10 + 81 = 280 + 81$

From the above nine examples, we can generalize the formula as follows:

When a number is between 11 and 19, the magic number is the ones digit of the number. The square of "the number" can then be calculated with the following formula:

Formula 3.1: (the number)² = (10 + magic number)² = (the number + magic number) x 10 + (magic number)²

Formula 3.1 is similar to **Formula 2.2** (see **Section 2.3**). The only difference is that 10 is used in Formula 3.1 instead of 100 is used in **Formula 2.2**.

We formally express **Formula 3.1** as follows. We call a number A and the magic number (ones digit) m, where A = 10 + m,

$$\text{Formula 3.1}: A^2 = (10 + m)^2 = (A + m) \times 10 + m^2$$

We use **Formula 3.1** to calculate 18² as follows: A = 18, m = 8, applying **Formula 3.1** as

$$18^2 = (18 + 8) \times 10 + 8^2 = 260 + 64 = 324$$

Can you remember 17² = 289? If you can, you don't need the rapid calculation. If you can't, then calculate it as follows:

The magic number is m = 7, so
17² = (17 + 7) x 10 + 7² = 240 + 49 = 289

If you want to know why Formula 3.1 is correct, please refer to Chapter 5 (Section 5.4) for the mathematical proof of the quick formula for the square of magic numbers.

In **Formula 3.1**, we first note the magic number (ones digit). Then in the second step, we calculate (the number + the magic number), and the third step calculates (the number + magic number) x 10, or add a 0 at the end. The fourth step calculates the square of the magic number, and the fifth step does the final addition, where there may be carrying.

The first step to work out the ones digit is not a calculation step. The second step is a simple addition. The third step is adding a 0 after a number, and it is not a calculation step. The fourth step calculates the square of the magic number (ones digit). The square of the number, the addition in the fifth step may have a carry, so it is a little difficult. After the above analysis, this Formula 3.1 also uses only three steps, but two of those steps are slightly more complicated. This is by far the most complicated calculation, but it is still 5 steps short of the 8 steps used in traditional two-digit multiplication.

From the above example, since we already know

11² = 121   12² = 144   13² = 169   14² = 196

And so on, applying Formula 3.1, we can calculate the square of 21 to 24 as follows:

The magic number of 21: m = 21 − 10 = 11
21^2 = (21 + **11**) x 10 + **11**^{2} = 32 x 10 + 121 = 320 + 121 = 441
The magic number of 22: m = 22 − 10 = 12
22^2 = (22 + **12**) x 10 + **12**^{2} = 34 x 10 + 144 = 340 + 144 = 484
The magic number of 23: m = 23 − 10 = 13
23^2 = (23 + **13**) x 10 + **13**^{2} = 36 x 10 + 169 = 360 + 169 = 529
The magic number of 24: m = 24 − 10 = 14
24^2 = (24 + **14**) x 10 + **14**^{2} = 38 x 10 + 196 = 380 + 196 = 576

The squares of 11 - 24 will be used frequently in later algorithms, so I recommend you memorize their squares, which will further speed up the calculation. If you occasionally forget, use the formula above to calculate them.

You may ask, what is the relationship between reciting and memorizing these common numbers and rapid calculation? We all memorize the nine-by-nine multiplication table instead of calculating it every time. Memory can definitely improve the calculation speed. That is, the more memory, the faster the calculation. However, memorizing the squares is difficult, so we can only remember a few. But in fact, if you can memorize the squares of these 13 (from 11 to 19 and from 21 to 24) numbers, it will be of great help to later learning. If it's too difficult for you, it doesn't matter, but the calculation speed will be a little slower.

When a certain number is between 11 - 24, we abstract Formula 3.1 into the following four steps:

Step 1: Calculate the magic number (the number - 10), the magic number is between 1 - 14;
Step 2: Calculate (the number + magic number) and multiply it by 10;
Step 3: Calculate the square of the magic number.
Step 4: Add the result of step 3 to the result of step 2.

Exercise 3.1: In the following exercises, please work out what the magic number is for each question, and then calculate the answer.
Calculate the square of the following numbers.

18^2 = 15^2 = 12^2 = 19^2 =

11^2 = 17^2 = 13^2 = 16^2 =

14^2 = 24^2 = 21^2 = 23^2 =

22^2 =

Or recite 11 - 24 squared.

Section 3.2: 61 - 74 Squared (difficulty 3*)

Assumed knowledge: It is assumed that the reader is proficient in calculating the square of 11 - 24 in Section 3.1 or can memorize the values of the square of 11 - 24. If you are not proficient, it is best to review the method in **Section 3.1** several times.

Please see the examples below:

$61^2 = 3721$
$62^2 = 3844$
$63^2 = 3969$
$64^2 = 4096$
$65^2 = 4225$
$66^2 = 4356$
$67^2 = 4489$
$68^2 = 4624$
$69^2 = 4761$

$71^2 = 5041$
$72^2 = 5184$
$73^2 = 5329$
$74^2 = 5476$

If the above examples are disassembled and expressed as follows, you can see the pattern.

$61^2 = (50 + \mathbf{11})^2 = (25 + \mathbf{11}) \times 100 + \mathbf{11}^2 = 36 \times 100 + 121 = 3600 + 121 = 3721$
$62^2 = (50 + \mathbf{12})^2 = (25 + \mathbf{12}) \times 100 + \mathbf{12}^2 = 37 \times 100 + 144 = 3700 + 144 = 3844$
$63^2 = (50 + \mathbf{13})^2 = (25 + \mathbf{13}) \times 100 + \mathbf{13}^2 = 38 \times 100 + 169 = 3800 + 169 = 3969$
$64^2 = (50 + \mathbf{14})^2 = (25 + \mathbf{14}) \times 100 + \mathbf{14}^2 = 39 \times 100 + 196 = 3900 + 196 = 4096$
$65^2 = (50 + \mathbf{15})^2 = (25 + \mathbf{15}) \times 100 + \mathbf{15}^2 = 40 \times 100 + 225 = 4000 + 225 = 4225$
$66^2 = (50 + \mathbf{16})^2 = (25 + \mathbf{16}) \times 100 + \mathbf{16}^2 = 41 \times 100 + 256 = 4100 + 256 = 4356$
$67^2 = (50 + \mathbf{17})^2 = (25 + \mathbf{17}) \times 100 + \mathbf{17}^2 = 42 \times 100 + 289 = 4200 + 289 = 4489$
$68^2 = (50 + \mathbf{18})^2 = (25 + \mathbf{18}) \times 100 + \mathbf{18}^2 = 43 \times 100 + 324 = 4300 + 324 = 4624$
$69^2 = (50 + \mathbf{19})^2 = (25 + \mathbf{19}) \times 100 + \mathbf{19}^2 = 44 \times 100 + 361 = 4400 + 361 = 4761$
$70^2 = (50 + \mathbf{20})^2 = (25 + \mathbf{20}) \times 100 + \mathbf{20}^2 = 45 \times 100 + 400 = 4500 + 400 = 4900$
$71^2 = (50 + \mathbf{21})^2 = (25 + \mathbf{21}) \times 100 + \mathbf{21}^2 = 46 \times 100 + 441 = 4600 + 441 = 5041$
$72^2 = (50 + \mathbf{22})^2 = (25 + \mathbf{22}) \times 100 + \mathbf{22}^2 = 47 \times 100 + 484 = 4700 + 484 = 5184$
$73^2 = (50 + \mathbf{23})^2 = (25 + \mathbf{23}) \times 100 + \mathbf{23}^2 = 48 \times 100 + 529 = 4800 + 529 = 5329$
$74^2 = (50 + \mathbf{24})^2 = (25 + \mathbf{24}) \times 100 + \mathbf{24}^2 = 49 \times 100 + 576 = 4900 + 576 = 5476$

Do you see the pattern? Yes, the algorithm for the square of 61 - 74 is the same as the algorithm in Section 2.1. If we define magic number = the number - 50, then the square of the number is calculated using Formula 2.1 (**see Section 2.1**).

Formula 2.1 can be interpreted as thus: the magic number is still the difference between "the number" and 50. The square of the number is part-1 plus part-2. The part-1 is 25 plus the magic number by timing 100, and the part-2 is the square of the magic number. As the magic number is greater than 10, the part-2 (square of the magic number) is three digits, so calculation of part-1 plus part-2 with carrying. Since the magic number for 61 - 74 is 11 – 24 respectively, you should remember what the square of 11 - 24 is. If you can't remember, please reread Section 3.1.

Now, we are using Formula 2.1 to explain as follows. For example, A = 63, the magic number of A is m = 63 – 50 = 13, so, $A^2 = 63^2 = (50 + 13)^2 = (25 + 13) \times 100 + 13^2 = 38 \times 100 + 169 = 3969$. The difference is that when calculating the square of 61 - 74, the magic number becomes larger, its square is greater than 100, the last two digits cannot be placed, and the hundreds digit has to be rounded up to the first two digits.

Here are some other examples:

$61^2 =?$

Since A = 61, the magic number of A is m = 61 – 50 = 11, and the magic number $m^2 = \mathbf{11}^2 = 121$, so:

$61^2 = (50 + \mathbf{11})^2 = (25 + \mathbf{11}) \times 100 + \mathbf{11}^2 = 36 \times 100 + 121 = 3600 + 121 = 3721$

$73^2 =?$

Since A = 73, the magic number of A is m = 73 – 50 = **23**, and the magic number $m^2 = \mathbf{23}^2 = 529$, so:

$73^2 = (50 + \mathbf{23})^2 = (25 + \mathbf{23}) \times 100 + \mathbf{23}^2 = 48 \times 100 + 529 = 4800 + 529 = 5329$

In fact, in this formula, we first calculate the magic number (not the ones digits). The second step calculates (25 + magic number). The third step calculates (25 + magic number) x 100, or we can add two 0s at the back of (25 + magic number). The fourth step calculates the square of the magic number. The fifth step calculates an addition.

The first two steps are easy. The third step is to add two 0s after a number, which cannot be counted as a step. Step 4 can be easy or difficult, because all magic numbers are less than 25. As long as you can remember the square of the 13 numbers in Section 3.1, Step 4 becomes easy. Otherwise, you must master Section 3.1. Formula 3.1, and then calculate the square of the magic number each time. In the fifth step, because the square of the magic number is greater than 100, the last two digits are not enough, so we need to calculate an addition with carrying, but addition is easy to calculate. After the above analysis, this Formula 2.1 also uses only 4 steps. Although two of the steps are slightly more complicated, it is still 4 steps less than the 8 steps of the traditional two-digit multiplication.

When a certain number is between 61 - 74, we abstract Formula 2.1 into the following four steps:

Step 1: Calculate the magic number (the number - 50), the magic number is between 11 - 24;
Step 2: Calculate (25 + magic number) multiplied by 100;
Step 3: Calculate the square of the magic number (between 121 and 576).
Step 4: Add the result of step 3 to the result of step 2.

If you want to calculate the square of 65, it is easier to use **Formula 1.1**.

Exercise 3.2: In the following exercises, please first calculate the magic number of each question, and then calculate the answer.

$66^2 =$ $62^2 =$ $63^2 =$ $64^2 =$

$68^2 =$ $71^2 =$ $73^2 =$ $65^2 =$

$67^2 =$ $69^2 =$ $61^2 =$ $74^2 =$

$72^2 =$

Section 3.3: 26 - 39 Squared (difficulty 4*)

Assumed knowledge: It is assumed that the reader is proficient in calculating the square of 11 - 24 in Section 3.1 or has memorized the values of the squares of 11 - 24, and understands the operation of negative numbers. Formula 2.1 will also be used in this section (see **Section 2.2**).

Please see the examples below:

$39^2 = 1521$
$38^2 = 1444$
$37^2 = 1369$
$36^2 = 1296$
$35^2 = 1225$

$34^2 = 1156$
$33^2 = 1089$
$32^2 = 1024$
$31^2 = 961$

$29^2 = 841$
$28^2 = 784$
$27^2 = 729$
$26^2 = 676$

If the above example are disassembled and expressed as follows, you can see the pattern:

$39^2 = (50 + (-11))^2 = (25 + (-11)) \times 100 + (-11)^2 = (25 - 11) \times 100 + 11^2 = 1400 + 121 = 1521$

$38^2 = (50 + (-12))^2 = (25 + (-12)) \times 100 + (-12)^2 = (25 - 12) \times 100 + 12^2 = 1300 + 144 = 1444$

$37^2 = (50 + (-13))^2 = (25 + (-13)) \times 100 + (-13)^2 = (25 - 13) \times 100 + 13^2 = 1200 + 169 = 1369$

$36^2 = (50 + (-14))^2 = (25 + (-14)) \times 100 + (-14)^2 = (25 - 14) \times 100 + 14^2 = 1100 + 196 = 1296$

$35^2 = (50 + (-15))^2 = (25 + (-15)) \times 100 + (-15)^2 = (25 - 15) \times 100 + 15^2 = 1000 + 225 = 1225$

$34^2 = (50 + (-16))^2 = (25 + (-16)) \times 100 + (-16)^2 = (25 - 16) \times 100 + 16^2 = 900 + 256 = 1156$

$33^2 = (50 + (-17))^2 = (25 + (-17)) \times 100 + (-17)^2 = (25 - 17) \times 100 + 17^2 = 800 + 289 = 1089$

$32^2 = (50 + (-18))^2 = (25 + (-18)) \times 100 + (-18)^2 = (25 - 18) \times 100 + 18^2 = 700 + 324 = 1024$

$31^2 = (50 + (-19))^2 = (25 + (-19)) \times 100 + (-19)^2 = (25 - 19) \times 100 + 19^2 = 600 + 361 = 961$

$30^2 = (50 + (-20))^2 = (25 + (-20)) \times 100 + (-20)^2 = (25 - 20) \times 100 + 20^2 = 500 + 400 = 900$

Or $30^2 = 900$

$29^2 = (50 + (-21))^2 = (25 + (-21)) \times 100 + (-21)^2 = (25 - 21) \times 100 + 21^2 = 400 + 441 = 841$

$28^2 = (50 + (-22))^2 = (25 + (-22)) \times 100 + (-22)^2 = (25 - 22) \times 100 + 22^2 = 300 + 484 = 784$

$27^2 = (50 + (-23))^2 = (25 + (-23)) \times 100 + (-23)^2 = (25 - 23) \times 100 + 23^2 = 200 + 529 = 729$

$26^2 = (50 + (-24))^2 = (25 + (-24)) \times 100 + (-24)^2 = (25 - 24) \times 100 + 24^2 =$

100 + 576 = 676

As you may have guessed, the algorithm for the square of 26 - 39 is the same as the algorithm in the previous section (section 3.2). If we define the magic number = the number - 50, then the square of the number uses Formula 2.1 (see Section 2.2). Repeat as follows:

Formula 2.1: (2-1): $A^2 = (50 + m)^2 = (25 + m) \times 100 + m^2$

A reminder that for 26 - 39, the magic number is negative. Formula 2.1 can be interpreted as: the magic number is still the difference between the number and 50. The square of the number are part-1 plus part-2. The part-1 is 25 plus the magic number by timing 100, and the part-2 is the square of the magic number. As the magic number is less than -10, the part-2 (square of the magic number) is three digits, so calculation of part-1 plus part-2 with carrying.

Since the magic numbers for 26 - 39 are -11 to -24, their squares are the same as 11 - 24 squared. Therefore, you should remember what the squares of 11 – 24 is. If you don't remember them well, please reread **Section 3.1**.

Here are some other examples:
What is 39^2?
Because the magic number of 39 is m = 39 – 50 = -11, the magic number $m^2 = (-11)^2 = 11^2 = 121$, so:

$39^2 = (50 + (-11))^2 = (25 + (-11)) \times 100 + (-11)^2 = 14 \times 100 + 121 = 1400 + 121 = 1521$

What is 27^2?
Because the magic number of 27 is m = 27 – 50 = -23, the magic number $m^2 = (-23)^2 = 23^2 = 529$, so:

$27^2 = (50 + (-23))^2 = (25 + (-23)) \times 100 + (-23)^2 = 2 \times 100 + 529 = 200 + 529 = 729$

In **Formula 2.1**, we calculate the magic number in the first step, (25 + magic number) in the second step, and (25 + magic number) x 100 in the third step, or add two 0s. The fourth step calculates the square of the magic number, and the fifth step calculates an addition. Because the magic number is negative, the second step is actually a subtraction. The third step cannot be counted as a calculation step. In the fourth step, it should be noted that the square of a negative number is equal to the square of a positive number. In the fifth step, because the square of the magic number is greater than 100, the last two digits are not enough, so an addition with carrying must be calculated, but after all, addition is easy to calculate. Thus, Formula 2.1 actually takes only 4 steps here.

When a certain number is between 39 and 26, we abstract Formula 2.1 into the following four steps:

Step 1: Calculate the magic number (the number - 50), the magic number is between (-11) - (-24);
Step 2: Calculate (25 + magic number) multiplied by 100;
Step 3: Calculate the square of the magic number (between 121 and 576).
Step 4: Add the result of step 3 to the result of step 2.

If you want to calculate the square of 35, it is easier to use Formula **1.1**.
You may notice that the same Formula 2.1 is used when we square 26 - 74. See Sections 2.1, 2.2, 3.2 and 3.3.

Exercise 3.3: In the following exercises, please calculate the magic number of each question first, and then calculate the answer.

$28^2 =$	$31^2 =$	$33^2 =$	$34^2 =$
$37^2 =$	$39^2 =$	$26^2 =$	$27^2 =$
$38^2 =$	$36^2 =$	$29^2 =$	$32^2 =$

Section 3.4: 111 - 124 Squared (difficulty 3*)

Assumed knowledge: It is assumed that the reader is proficient in calculating the squares of 11 - 24 in Section 3.1 or has memorized the values of the squares of 11 - 24. Formula 2.2 will also be used in this section (see **Section 2.3**).

Please see the examples below:

$111^2 = 12321$
$112^2 = 12544$
$113^2 = 12769$
$114^2 = 12996$
$115^2 = 13225$
$116^2 = 13456$
$117^2 = 13689$
$118^2 = 13924$

$119^2 = 14161$

$121^2 = 14641$
$122^2 = 14884$
$123^2 = 15129$
$124^2 = 15376$

If the above examples are disassembled and expressed as follows, you can see the pattern:

$111^2 = (100 + \mathbf{11})^2 = (111 + \mathbf{11}) \times 100 + \mathbf{11}^2 = 122 \times 100 + 121 = 12200 + 121 = 12321$
$112^2 = (100 + \mathbf{12})^2 = (112 + \mathbf{12}) \times 100 + \mathbf{12}^2 = 124 \times 100 + 144 = 12400 + 144 = 12544$
$113^2 = (100 + \mathbf{13})^2 = (113 + \mathbf{13}) \times 100 + \mathbf{13}^2 = 126 \times 100 + 169 = 12600 + 169 = 12769$
$114^2 = (100 + \mathbf{14})^2 = (114 + \mathbf{14}) \times 100 + \mathbf{14}^2 = 128 \times 100 + 196 = 12800 + 196 = 12996$
$115^2 = (100 + \mathbf{15})^2 = (115 + \mathbf{15}) \times 100 + \mathbf{15}^2 = 130 \times 100 + 225 = 13000 + 225 = 13225$
$116^2 = (100 + \mathbf{16})^2 = (116 + \mathbf{16}) \times 100 + \mathbf{16}^2 = 132 \times 100 + 256 = 13200 + 256 = 13456$
$117^2 = (100 + \mathbf{17})^2 = (117 + \mathbf{17}) \times 100 + \mathbf{17}^2 = 134 \times 100 + 289 = 13400 + 289 = 13689$
$118^2 = (100 + \mathbf{18})^2 = (118 + \mathbf{18}) \times 100 + \mathbf{18}^2 = 136 \times 100 + 324 = 13600 + 324 = 13924$
$119^2 = (100 + \mathbf{19})^2 = (119 + \mathbf{19}) \times 100 + \mathbf{19}^2 = 138 \times 100 + 361 = 13800 + 361 = 14161$
$120^2 = (100 + \mathbf{20})^2 = (120 + \mathbf{20}) \times 100 + \mathbf{20}^2 = 140 \times 100 + 400 = 14000 + 400 = 14400$

Or $120^2 = 12^2 \times 100 = 144 \times 100 = 14400$

$121^2 = (100 + \mathbf{21})^2 = (121 + \mathbf{21}) \times 100 + \mathbf{21}^2 = 142 \times 100 + 441 = 14200 + 441 = 14641$
$122^2 = (100 + \mathbf{22})^2 = (122 + \mathbf{22}) \times 100 + \mathbf{22}^2 = 144 \times 100 + 484 = 14400 + 484 = 14884$
$123^2 = (100 + \mathbf{23})^2 = (123 + \mathbf{23}) \times 100 + \mathbf{23}^2 = 146 \times 100 + 529 = 14600 + 529 = 15129$
$124^2 = (100 + \mathbf{24})^2 = (124 + \mathbf{24}) \times 100 + \mathbf{24}^2 = 148 \times 100 + 576 = 14800 + 576 = 15376$

The algorithm for computing the square of 101 – 109 was introduced in **Section 2.3**. The formula is as follows:

$$\textbf{Formula 2-2:}\ A^2 = (100 + m)^2 = (A + m) \times 100 + m^2$$

Formula 2.2 is also suitable for calculating the squares of 111 – 124. The difference is that the magic numbers range from 1 - 9 to 11 - 24. The square of the number are part-1 plus part-2. The part-1 is the number itself plus the magic number by timing 100, and the part-2 is the square of the magic number. As the magic number is greater than 10, the part-2 (square of the magic number) is greater than 100 (three digits), so calculation of part-1 plus part-2 with carrying.

Now, we use Formula 2.2 to explain as follows. The magic number for A = 113 is m = 113 – 100 = 13, and $13^2 = 169$, so, $113^2 = (100 + 13)^2 = (113 + 13) \times 100 + 13^2 = 126 \times 100 + 169 = 12769$.

Here are some other examples:

$111^2 = ?$

Because the magic number m = 111 – 100 = 11 for A = 111, the magic number $m^2 = 11^2 = 121$, so:

$111^2 = (100 + \mathbf{11})^2 = (111 + \mathbf{11}) \times 100 + \mathbf{11}^2 = 122 \times 100 + 121 = 12200 + 121 = 12321$

$118^2 = ?$

Because the magic number m = 118 – 100 = 18 for A = 118, the magic number $m^2 = \mathbf{18}^2 = 324$, so:

$118^2 = (100 + \mathbf{18})^2 = (118 + \mathbf{18}) \times 100 + \mathbf{18}^2 = 136 \times 100 + 324 = 13600 + 324 = 13924$

In **Formula 2.2**, we calculate the magic number in the first step (not the ones digit). In the second step, we calculate (the number itself + the magic number). In the third step we calculate (the number itself + the magic number) x 100, or add two 0s after it. In the fourth step, we calculate the square of the magic number. In the fifth step, we calculate an addition. The first three steps are simple to calculate. The fourth step can be easy or difficult (see the analysis in Section 3.2). The fifth step is to calculate an addition with carrying, which is also easy to calculate. **Formula 2.2** actually takes only 4 steps here.

When a certain number is between 111 - 124, we abstract **Formula 2.2** into the following four steps:

Step 1: Calculate the magic number (the number - 100), the magic number is between 11 - 24;
Step 2: Calculate (the number + magic number) and multiply it by 100;
Step 3: Calculate the square of the magic number (between 121 and 576).
Step 4: Add the result of step 3 to the result of step 2.

115^2 has two algorithms. The Formula 1.1 introduced in Section 1.2 is: $115^2 = (11 \times (11 + 1)) \times 100 + 25 = 13225$, the magic number is 11, which consists of 2 steps of calculation. The Formula 2.2 introduced in this section 3.4 is: $115^2 = (115 + 15) \times 100 + 15^2 = 13225$, with the magic number being 15. This formula includes 3 steps of calculation. Both algorithms give the same result, and it is easier to use the **Formula 2.2**.

You should memorize the squared values of 11 – 24 before doing the following exercises.

Exercise 3.4: In each of the following exercises, please calculate the magic number, and then calculate the answer.

$124^2 =$ $116^2 =$ $119^2 =$ $123^2 =$

$115^2 =$ $117^2 =$ $113^2 =$ $122^2 =$

$121^2 =$ $111^2 =$ $118^2 =$ $112^2 =$

$114^2 =$

Section 3.5: 76 - 89 Squared (difficulty 4*)

Assumed knowledge: It is assumed that the reader is proficient in calculating the squares of 11 - 24 in Section 3.1 or has memorized the values of the squares of 11 - 24. Formula 2.2 will also be used in this section (see **Section 2.3**).

Please see the examples below:

$89^2 = 7921$
$88^2 = 7744$
$87^2 = 7569$
$86^2 = 7396$
$85^2 = 7225$
$84^2 = 7056$

$83^2 = 6889$
$82^2 = 6724$
$81^2 = 6561$

$79^2 = 6241$
$78^2 = 6084$
$77^2 = 5929$
$76^2 = 5776$

If the above examples are disassembled and expressed as follows, you can see the pattern:

$89^2 = (100 + (-11))^2 = (89 + (-11)) \times 100 + (-11)^2 = 78 \times 100 + 121 = 7921$

Note here that $(-11)^2 = 11^2 = 121$
And so on:

$88^2 = (100 + (-12))^2 = (88 + (-12)) \times 100 + (-12)^2 = 76 \times 100 + 144 = 7744$
$87^2 = (100 + (-13))^2 = (87 + (-13)) \times 100 + (-13)^2 = 74 \times 100 + 169 = 7569$
$86^2 = (100 + (-14))^2 = (86 + (-14)) \times 100 + (-14)^2 = 72 \times 100 + 196 = 7396$
$85^2 = (100 + (-15))^2 = (85 + (-15)) \times 100 + (-15)^2 = 70 \times 100 + 225 = 7225$
$84^2 = (100 + (-16))^2 = (84 + (-16)) \times 100 + (-16)^2 = 68 \times 100 + 256 = 7056$
$83^2 = (100 + (-17))^2 = (83 + (-17)) \times 100 + (-17)^2 = 66 \times 100 + 289 = 6889$
$82^2 = (100 + (-18))^2 = (82 + (-18)) \times 100 + (-18)^2 = 64 \times 100 + 324 = 6724$
$81^2 = (100 + (-19))^2 = (81 + (-19)) \times 100 + (-19)^2 = 62 \times 100 + 361 = 6561$
$80^2 = (100 + (-20))^2 = (80 + (-20)) \times 100 + (-20)^2 = 60 \times 100 + 400 = 6400$

Or $80^2 = 8^2 \times 100 = 64 \times 100 = 6400$

$79^2 = (100 + (-21))^2 = (79 + (-21)) \times 100 + (-21)^2 = 58 \times 100 + 441 = 6241$
$78^2 = (100 + (-22))^2 = (78 + (-22)) \times 100 + (-22)^2 = 56 \times 100 + 484 = 6084$
$77^2 = (100 + (-23))^2 = (77 + (-23)) \times 100 + (-23)^2 = 54 \times 100 + 529 = 5929$
$76^2 = (100 + (-24))^2 = (76 + (-24)) \times 100 + (-24)^2 = 52 \times 100 + 576 = 5776$

Section 2.4 presents the algorithm for computing the squares of 99–91, and this section is an extension of it. The magic number is still the difference between the number and 100. This section extends the range of magic numbers from -1 to -9, to -11 to -24. The difficulty is remembering the squares of 11–24.

Using **Formula 2.2**, the calculation rule is as follows: For the squares of 89 - 76, the magic number is the difference between the number and 100. The square of the number are part-1 plus part-2. The part-1 is the number itself plus the magic number and then multiplying by 100, and the part-2 is the square of the magic number. As the magic number is less than -10, the part-2 (square of the magic number) is greater than 100 (three digits), so calculation of part-1 plus part-2 with carrying.

An example is as follows:

$89^2 = ?$

Since the magic number m = 89 − 100 = -11 for A = 89, the magic number $m^2 = (-11)^2 = 121$, so
$89^2 = (100 + (-11))^2 = (89 + (-11)) \times 100 + (-11)^2 = 78 \times 100 + 121 = 7921$

$78^2 = ?$

Because the magic number m = 78 − 100 = -22 for A = 78, the magic number $m^2 = (-22)^2 = 484$, so

$78^2 = (100 + (-22))^2 = (78 + (-22)) \times 100 + (-22)^2 = 56 \times 100 + 484 = 6084$

In **Formula 2.2**, the first step calculates the magic number, the second step calculates (the number + magic number), the third step calculates (the number + magic number) × 100, or add two 0s, the fourth step calculates the square of the magic number, and the fifth step calculates an addition. Because the magic number is negative, the second step is actually a subtraction. The third step is not computationally intensive. The fourth step is to note that the square of a negative number is equal to the square of a positive number. The fifth step is to compute an addition with carrying. Formula 2.2 actually takes only 4 steps here.

When a certain number is between 89 - 76, we abstract Formula 2.2 into the following four steps:

Step 1: Calculate the magic number (the number - 100), the magic number is between (-11) - (-24);
Step 2: Calculate (the number + magic number) and multiply it by 100;
Step 3: Calculate the square of the magic number (between 121 and 576).
Step 4: Add the result of step 3 to the result of step 2.

If you want to calculate the square of 85, it is easier to use **Formula 1.1**.

Exercise 3.5: In the following exercises, please calculate the magic number of each question first, and then calculate the result.

$76^2 =$	$79^2 =$	$87^2 =$	$83^2 =$
$77^2 =$	$78^2 =$	$82^2 =$	$81^2 =$
$86^2 =$	$89^2 =$	$88^2 =$	$84^2 =$

Section 3.6: Summary (difficulty 4*)

Chapters 2 and 3 describe algorithms for squaring, from simple to complex, from no-carrying to carrying. I recommend that you memorize the squares of 11-24 in order to calculate the squares of 1-125 easily.

It is important to remember that when you find (or calculate) the magic number, forget the original number, because the subsequent calculations are all based on the magic number.

For better understanding, let's take two more examples:

$39^2 = ?$

Its magic number m = 39 - 50 = -11, so, $39^2 = (25 + (-11)) \times 100 + (-11)^2 =$ 1400 + 121 = 1521

$87^2 = ?$

Its magic number m = 87 - 100 = -13, so, $87^2 = (87 + (-13)) \times 100 + (-13)^2$ = 7400 + 169 = 7569

Congratulations, you have mastered a lot of speed calculation skills! Write out a few examples, look for patterns, and then verify whether you can go further. For example, can you find an easy algorithm to calculate the squares of 126-200? If you're curious, continue to Chapter 4.

To recap, we were able to mentally calculate the squares of 1 - 125 using only two neat formulas (Formula 2.1 and Formula 2.2). For the sake of memory, I will summarize the core of these two chapters:

When a number is around 50 (26 - 74), the number = 50 + magic number, and -25 < magic number < 25. The magic number of the number = the number – 50. When a number is greater than 50, the magic number is positive; when a number is less than 50, the magic number is negative. (the number)² can be expressed by the formula as follows:

Formula 2.1: (the number)² = (50 + magic number)² = (25 + magic number) x 100 + (magic number)²

Note that (25 + magic number) x 100 means that (25 + magic number) takes the first two digits and (magic number)² takes the last two digits. If the magic number is greater than 10, or the square of the magic number is greater than 100, the (magic number)² has a carry.

It can also be expressed formally as A = 50 + m, or m = A − 50, 25 < A < 75, so -25 < m < 25,

Formula 2.1: A² = (50 + m)² = (25 + m) x 100 + m²

Since |m| < 25, a square calculation around 50 reduces the dimensionality to an addition and a square less than 25.

When a certain number is around 100 (76 - 124), the number = 100 + magic number. When the number is greater than 100, the magic number is positive; when the number is less than 100, the magic number is negative. (the number)² can thus be expressed by the formula as follows:

Formula 2.2: (the number)² = (100 + magic number)² = (the number + magic number) x 100 + (magic number)²

Note that (the number + magic number) x 100 means it occupies the first two (or three) digits. (Magic number)² occupies the last two digits. If the magic number is greater than 10, or the square of the magic number is greater than 100, the (magic number)² has a carry.

It can also be expressed formally as A = 100 + m, or m = A − 100, 75 < A < 125, so -25 < m < 25,

Formula 2.2: A² = (100 + m)² = (A + m) x 100 + m²

Since |m| < 25, the squaring calculation around 100 is reduced to an addition and a squaring calculation of a number less than 25.

So, the squaring of any number less than 125 is reduced to an addition and the squaring of a number less than 25. In other words, as long as you can calculate the squares of numbers within 25, you can quickly calculate the square of any number within 125.

Let's now summarize the square rapid calculation for 1 - 125. With only 125 numbers, we can group the list as follows:

Group 1: The Square of 52 numbers:

	$50^2 = 2500$		$100^2 = 10000$	
$1^2 = 1$	$51^2 = 2601$	$49^2 = 2401$	$101^2 = 10201$	$99^2 = 9801$
$2^2 = 4$	$52^2 = 2704$	$48^2 = 2304$	$102^2 = 10404$	$98^2 = 9604$
$3^2 = 9$	$53^2 = 2809$	$47^2 = 2209$	$103^2 = 10609$	$97^2 = 9409$
$4^2 = 16$	$54^2 = 2916$	$46^2 = 2116$	$104^2 = 10816$	$96^2 = 9216$
$5^2 = 25$	$55^2 = 3025$	$45^2 = 2025$	$105^2 = 11025$	$95^2 = 9025$
$6^2 = 36$	$56^2 = 3136$	$44^2 = 1936$	$106^2 = 11236$	$94^2 = 8836$
$7^2 = 49$	$57^2 = 3249$	$43^2 = 1849$	$107^2 = 11449$	$93^2 = 8649$
$8^2 = 64$	$58^2 = 3364$	$42^2 = 1764$	$108^2 = 11664$	$92^2 = 8464$
$9^2 = 81$	$59^2 = 3481$	$41^2 = 1681$	$109^2 = 11881$	$91^2 = 8281$
$10^2 = 100$	$60^2 = 3600$	$40^2 = 1600$	$110^2 = 12100$	$90^2 = 8100$

Let's analyze the complexity of these computations.

The first column is calculated in one step (see **Chapter 1**).

The second column is calculated in two steps. The first step is to calculate 25 plus the ones digit of the number and place it in the first two digits, and the second step is to calculate the square of the ones digit and place it in the last two digits (see **Section 2.1**, **Formula 2.1**).

The third column is calculated in three steps. The first step is to calculate the magic number, the second step is to add 25 to the magic number and place it in the first two digits, and the third step is to calculate the square of the magic number and place it in the last two digits (see **Section 2.2**, **Formula 2.1**).

The fourth column is similar to the second column, and only takes two calculations. The first step is to calculate the number itself plus the ones digit and place it in the first three digits. The second step is to calculate the square of the ones digit and place it in the last two digits (see **Section 2.3**, **Formula 2.2**).

The fifth column is similar to the third column, and only takes three calculations. The first step is to calculate the magic number. The second step is to add the magic number to the number itself and put it in the first two digits. The third step is to calculate the square of the magic number and put it in the last two digits (see **Section 2.4**, **Formula 2.2**),

Look at the rows and you will see that the last two digits of each row are the same.

Group 2: The Square of 50 Numbers.

$11^2 = 121$	$61^2 = 3721$	$39^2 = 1521$	$111^2 = 12321$	$89^2 = 7921$
$12^2 = 144$	$62^2 = 3844$	$38^2 = 1444$	$112^2 = 12544$	$88^2 = 7744$
$13^2 = 169$	$63^2 = 3969$	$37^2 = 1369$	$113^2 = 12769$	$87^2 = 7569$
$14^2 = 196$	$64^2 = 4096$	$36^2 = 1296$	$114^2 = 12996$	$86^2 = 7396$
$15^2 = 225$	$65^2 = 4225$	$35^2 = 1225$	$115^2 = 13025$	$85^2 = 7225$
$16^2 = 256$	$66^2 = 4356$	$34^2 = 1156$	$116^2 = 13456$	$84^2 = 7056$
$17^2 = 289$	$67^2 = 4489$	$33^2 = 1089$	$117^2 = 13689$	$83^2 = 6889$
$18^2 = 324$	$68^2 = 4624$	$32^2 = 1024$	$118^2 = 13924$	$82^2 = 6724$
$19^2 = 361$	$69^2 = 4761$	$31^2 = 961$	$119^2 = 14161$	$81^2 = 6561$
$20^2 = 400$	$70^2 = 4900$	$30^2 = 900$	$120^2 = 14400$	$80^2 = 6400$

Let's analyze the complexity of the computations.

The first column only needs three steps to calculate. The first step is to add the ones digit to the number, followed by a zero. The second step is to calculate the square of the ones digit. The third step is addition (see **Section 3.1**, **Formula 3.1**).

The second column requires a four-step calculation. The first step is to calculate the magic number. The second step is to calculate 25 plus the magic number, followed by two 0's. The third step is to calculate the square of the magic number (the square of the first column). The fourth step is addition (see **Section 3.2**, **Formula 2.1**).

The calculation of the third column is similar to the calculation of the second column (see **Section 3.3**, **Formula 2.1**).

The fourth column requires only four calculations. The first step is to calculate the magic number. The second step is to add the number and the magic number and place them in the first three digits. The third step is to calculate the square of the magic number (the square of the first column). The fourth step is to do an addition (see **Section 3.4**, **Formula 2.2**).

The fifth column is similar to the fourth column. The first step is to calculate the magic number, the second step is to calculate the number plus the magic number and put it in the first two digits, the third step is to calculate the square of the magic number (the square of the first column), and the fourth step is to add (see **Section 3.5**, **Formula 2.2**).

Look at the rows, and you will see that the last two digits of each row are the same.

Group 3: The Square of 23 Numbers.

$21^2 = 441$ $71^2 = 5041$ $29^2 = 541$ $121^2 = 14641$ $79^2 = 6241$

$22^2 = 484$ $72^2 = 5184$ $28^2 = 684$ $122^2 = 14884$ $78^2 = 6084$

$23^2 = 529$ $73^2 = 5329$ $27^2 = 829$ $123^2 = 15129$ $77^2 = 5929$

$24^2 = 576$ $74^2 = 5476$ $26^2 = 976$ $124^2 = 15376$ $76^2 = 5776$

$25^2 = 625$ $75^2 = 5625$ $125^2 = 15625$

This group is the most complex. However, just remember the squares of the four numbers (21 – 24) in the first column, Formula 2.1 for the second and third columns, and Formula 2.2 for the fourth and fifth columns in order to calculate.

Look at the rows, and you will also note that the last two digits of each row are the same.

So far, we have learned to mentally calculate the squares of numbers from 1 to 125.

It can also be seen from the above examples that the order of numbers is important for identifying regularities.

Comprehensive Exercises in this chapter:

$15^2 =$ $65^2 =$ $35^2 =$ $115^2 =$ $85^2 =$

$16^2 =$ $66^2 =$ $36^2 =$ $116^2 =$ $86^2 =$

$18^2 =$ $68^2 =$ $38^2 =$ $118^2 =$ $88^2 =$

$14^2 =$ $64^2 =$ $34^2 =$ $114^2 =$ $84^2 =$

$12^2 =$ $62^2 =$ $32^2 =$ $112^2 =$ $82^2 =$

$11^2 =$ $61^2 =$ $31^2 =$ $111^2 =$ $81^2 =$

$17^2 =$ $67^2 =$ $37^2 =$ $117^2 =$ $87^2 =$

$13^2 =$ $63^2 =$ $33^2 =$ $113^2 =$ $83^2 =$

$19^2 =$ $69^2 =$ $39^2 =$ $119^2 =$ $89^2 =$

$21^2 =$ $71^2 =$ $29^2 =$ $121^2 =$ $79^2 =$

$22^2 =$ $72^2 =$ $28^2 =$ $122^2 =$ $78^2 =$

$23^2 =$ $73^2 =$ $27^2 =$ $123^2 =$ $77^2 =$

$24^2 =$ $74^2 =$ $26^2 =$ $124^2 =$ $76^2 =$

CHAPTER 4: THE SQUARES OF 101 TO 999 (OPTIONAL READING)

Section 4.1: 101 - 199 Squared (difficulty 4*)

Assumed knowledge: It is assumed that the reader can mentally calculate the square of any number up to 100. If you still can't do it, please review the previous content until you are proficient in calculating the square of any number within 100, and then read on.

When we calculate the squares of 101 - 199, the magic number is still the difference between a certain number and 100. In other words, a certain number = 100 + magic number. This time the magic number should be in the range of 1 – 99. We've learned how to square the magic number (1 - 99). Thus, we can still use Formula 2.2 to calculate the squares of the numbers 101 - 199.

Formula 2.2: (the number) ^{2} = (100 + magic number) ^{2} = (the number + magic number) x 100 + (magic number) ^{2}

Once you can quickly calculate the squares of 1 – 99, which is (magic number) ^{2}, squaring 101 - 199 becomes simple. However, considering that addition is more complicated to calculate, this section is considered optional reading. If you find this section difficult, you can skip it.

Let's explain with a few examples:

142^2 = ?
The magic number of 142 is 142 – 100 = **42**, 42^2 = 1764 (using Formula 2.1), so
142^2 = (142 + **42**) x 100 + 42^2 = 18400 + 1764 = 20164

And so on:

134^2 = (134 + **34**) x 100 + 34^2 = 16800 + 1156 = 17956
157^2 = (157 + **57**) x 100 + 57^2 = 21400 + 3249 = 24649
163^2 = (163 + **63**) x 100 + 63^2 = 22600 + 3969 = 26569
171^2 = (171 + **71**) x 100 + 71^2 = 24200 + 5041 = 29241
189^2 = (189 + **89**) x 100 + 89^2 = 27800 + 7921 = 35721
195^2 = (195 + **95**) x 100 + 95^2 = 29000 + 9025 = 38025

This is kind of hard. Let's verify it with a special case:
$150^2 = 22500$, the magic number is $150 - 100 = 50$. The algorithm is as follows:
$150^2 = (150 + \mathbf{50}) \times 100 + \mathbf{50}^2 = 20000 + 2500 = 22500$

Although the above formula is simple, it takes a lot of practice to make perfect.

Exercise 4.1: In the following exercises, please calculate the magic number of each question first, and then calculate the answer:

$127^2 =$	$129^2 =$	$132^2 =$	$137^2 =$	$142^2 =$
$146^2 =$	$153^2 =$	$159^2 =$	$161^2 =$	$168^2 =$
$175^2 =$	$171^2 =$	$188^2 =$	$186^2 =$	$192^2 =$
$197^2 =$	$194^2 =$	$143^2 =$	$184^2 =$	$163^2 =$
$157^2 =$	$126^2 =$	$173^2 =$	$174^2 =$	$129^2 =$

Section 4.2: 201 – 999 Squared (difficulty 5*)

We formally express a number as $A = m_1 * 100 + m_2$, $200 < A < 1000$, so $1 < m_1 < 10$, $0 < m_2 < 100$

Formula 4.1: $A^2 = (m_1 * 100 + m_2)^2 = m_1 * (A + m_2) \times 100 + m_2^2$

The proof of Formula 4.1 is given in Section 5.5.
When $m_1 = 1$, Formula 4.1: $A^2 = (m_1 * 100 + m_2)^2 = m_1 * (A + m_2) \times 100 + m_2^2 = (A + m_2) \times 100 + m_2^2$ degenerates to Formula 2.2. In other words, Formula 2.2 is just a special case of Formula 4.1.

Formula 4.1 looks complicated at first glance, but it is actually quite simple. There are only four simple steps of calculation:

Step 1: $(A + m_2)$ is an addition of three-digits and two-digits. Addition is easy to do with mental arithmetic.

Step 2: $m_1 * (A + m_2) \times 100$ is a one-digit number multiplied by a three-digit or four-digit number, with two 0s added after the calculation result. It is not difficult to do with mental arithmetic.

Step 3: m_2^2 is a square calculation within 100, which we have already learned in previous chapters.

Step 4: Do another addition and add the calculation result of the second step to the calculation result of the third step.

We now give a few examples to calculate the squares of 201-999 using **Formula 4.1**.

$406^2 = ?$ where $m_1 = 4$, $m_2 = 6$,

$406^2 = 4 \times (406 + 6) \times 100 + 6^2 = 4 \times 412 \times 100 + 36 = 164800 + 36 = 164836$

$298^2 = ?$ where $m_1 = 2$, $m_2 = 98$,

$298^2 = 2 \times (298 + 98) \times 100 + 98^2 = 2 \times 396 \times 100 + 9604 = 79200 + 9604 = 88804$

$357^2 = ?$ where $m_1 = 3$, $m_2 = 57$,

$357^2 = 3 \times (357 + 57) \times 100 + 57^2 = 3 \times 414 \times 100 + 3249 = 124200 + 3249 = 127449$

$906^2 = ?$ where $m_1 = 9$, $m_2 = 6$,

$906^2 = 9 \times (906 + 6) \times 100 + 6^2 = 9 \times 912 \times 100 + 36 = 820800 + 36 = 820836$

$898^2 = ?$ where $m_1 = 8$, $m_2 = 98$,

$898^2 = 8 \times (898 + 98) \times 100 + 98^2 = 8 \times 996 \times 100 + 9604 = 796800 + 9604 = 806404$

$757^2 = ?$ where $m_1 = 7$, $m_2 = 57$,

$757^2 = 7 \times (757 + 57) \times 100 + 57^2 = 7 \times 814 \times 100 + 3249 = 569800 + 3249 = 573049$

Although it may seem difficult, it can be mastered. Without **Formula 4.1**, it would be difficult to mentally square numbers above six digits. As far as I know, Formula 4.1 is the least complex formula for mentally calculating the square of any three-digit number. Reduce the square of three digits, to one digit multiplied by multiple digits and two additions. **I call Formula 4.1 the "Chengqi Quick Calculation Method".**

Exercise 4.2: In the following exercises, please first calculate the two magic numbers of each question, and then calculate the answer:

$227^2 =$	$229^2 =$	$232^2 =$	$237^2 =$	$242^2 =$
$346^2 =$	$353^2 =$	$359^2 =$	$361^2 =$	$368^2 =$
$475^2 =$	$471^2 =$	$488^2 =$	$486^2 =$	$492^2 =$

CHAPTER 5: MATHEMATICAL PROOF OF THE MAGIC SQUARE QUICK FORMULA

This chapter gives mathematical proofs for all the quick formulas used for magic square number calculations.

Section 5.1: Proof of the Magic Square Quick Formula 1.1 with Ones Digit of 5 (difficulty 5*)

When the ones digit of a certain number A is 5 and the tens digit is any number, $A = 10 * m + 5$, and m is the magic number. The formula for calculating the square of A is (see Section 1.2)

Formula 1.1: $A^2 = (10 * m + 5)^2 = m * (m + 1) \times 100 + 25$

(Note: In the following proofs, when the two multipliers have letters, we use "*" for the multiplication sign, otherwise, we use "x" for the multiplication sign).

The proof is as follows:
We know: $A^2 - m^2 = (A - m) * (A + m)$
So $\quad A^2 = A^2 - 5^2 + 5^2$
$\quad\quad\quad = (A - 5) * (A + 5) + 5^2$
$\quad\quad\quad = (10 * m + 5 - 5) * (10 * m + 5 + 5) + 5^2$
$\quad\quad\quad = 10 * m * (10 * m + 10) + 5^2$
$\quad\quad\quad = 10 * m * 10 * (m + 1) + 5^2$
$\quad\quad\quad = m * (m + 1) \times 100 + 25$
So, we prove the **Formula 1.1:** $A^2 = m * (m + 1) \times 100 + 25$

Section 5.2: Proof of the Magic Square Quick Formula 2.1 for Numbers around 50 (difficulty 5*)

When A is between 26 and 74, A can be represented as $A = 50 + m$, or $m = A - 50$, where $25 < A < 75$, $-25 < m < 25$, m is the magic number. When the number A is greater than 50, m is a positive number. When the number A is less than 50, m is negative. (See Section 2.1, Section 2.2, Section 3.2, and Section 3.3). A^2 can be calculated with Formula 2.1.

Formula 2.1: $A^2 = (50 + m)^2 = (25 + m) \times 100 + m^2$

The proof is as follows:
We know: $A^2 - m^2 = (A - m) * (A + m)$
So, the **Formula 2.1**: $A^2 = A^2 - m^2 + m^2$
$$= (A - m) * (A + m) + m^2$$
$$= (50 + m - m) * (50 + m + m) + m^2$$
$$= 50 * (50 + 2*m) + m^2$$
$$= 2500 + 100 * m + m^2$$
$$= (25 + m) \times 100 + m^2$$
So, we prove the **Formula 2.1**: $A^2 = (25 + m) \times 100 + m^2$

Section 5.3: Proof of the Magic Square Quick Formula 2.2 for Numbers around 100 (difficulty 5*)

When A is between 76 and 124, A can be represented as $A = 100 + m$, or $m = A - 100$ where $75 < A < 125$, $-25 < m < 25$, m is the magic number. When this number A is greater than 100, m is a positive number. When this number A is less than 100, m is negative (see **Section 2.3**, **Section 2.4**, **Section 3.4**, **Section 3.5**, **Section 4.1**). Calculate A^2 by **Formula 2.2**:

Formula 2.2: $A^2 = (100 + m)^2 = (A + m) \times 100 + m^2$

The proof is as follows:
We know: $A^2 - m^2 = (A - m) * (A + m)$
$$A^2 = A^2 - m^2 + m^2$$
$$= (A - m) * (A + m) + m^2$$
$$= (100 + m - m) * (A + m) + m^2$$
$$= (A + m) \times 100 + m^2$$
So, we prove the **Formula 2.2**: $A^2 = (A + m) \times 100 + m^2$

Reader's Note: **Formula 2.2** also applies when A is between 101 – 199. In this case, the magic number is $0 < m < 100$ (see Section 4.1).

Section 5.4: Proof of the Magic Square Quick Formula 3.1 for 11 - 24 (difficulty 5*)

When A is between 11 and 24, A can be represented as $A = 10 + m$, or $m = A - 10$, where $10 < A < 25$, $0 < m < 15$, m is the magic number (see Section 3.1). Calculate A^2 using **Formula 3.1**:

Formula 3.1: $A^2 = (10 + m)^2 = (A + m) \times 10 + m^2$

The proof is as follows:
We know: $A^2 - m^2 = (A - m) * (A + m)$
So $A^2 = A^2 - m^2 + m^2$
$= (A - m) * (A + m) + m^2$
$= (10 + m - m) * (A + m) + m^2$
$= (A + m) \times 10 + m^2$

So, we prove the **Formula 3.1**: $A^2 = (A + m) \times 10 + m^2$

Section 5.5: Proof of Magic Square Quick Formula 4.1 for 100 - 1000 (difficulty 5*)

When a number A is between 101 and 999, A can be expressed as $A = m_1 * 100 + m_2$, so $0 < m_1 < 10$ and $0 < m_2 < 100$ (see **Section 4.2**).

Formula 4.1: $A^2 = (m_1 * 100 + m_2)^2 = m_1 * (A + m_2) \times 100 + m_2^2$

Now let's prove **Formula 4.1**.
$A^2 = A * (m_1 * 100 + m_2)$
$= A * m_1 * 100 + A * m_2$
$= A * m_1 * 100 + (m_1 * 100 + m_2) * m_2$
$= A * m_1 * 100 + m_1 * m_2 * 100 + m_2^2$
$= m_1 * (A + m_2) \times 100 + m_2^2$

So, we prove the **Formula 4.1**: $A^2 = m_1 * (A + m_2) \times 100 + m_2^2$

PART II
MAGIC MULTIPLICATION SPEED ALGORITHM

This part is divided into three chapters (Chapter 6 to Chapter 8).

Chapter 6 presents a universal rapid calculation formula for multiplication, which is based entirely on the rapid calculation of squares. Use this formula to quickly calculate the product of any two numbers. You can use this general rapid calculation formula only if you are proficient in the square rapid calculation used in Chapter 2 and Chapter 3 (quickly calculating squares within 125). This chapter is the essence of this book.

Chapter 7 explains some special multiplication formulas. These formulas are only suitable for some special cases. The application is narrow, but the calculation speed is relatively fast. In these special cases, we prefer to use these formulas. When no special multiplication quick formula is available, we can use the universal formula from Chapter 6.

Chapter 8 proves the correctness of the rapid calculation formula of magic multiplication with a strict mathematical method.

CHAPTER 6: MAGIC MULTIPLICATION OF ANY TWO NUMBERS (UNIVERSAL FORMULA)

If you are not proficient in mentally calculating the squares of 1 - 99, please review Chapters 2 and 3 first, and then read on until you are proficient.

Section 6.1: Universal Formula for Magic Multiplication (difficulty 4*)

We know that the squares of 1 - 99 only have 99 results, but the multiplication involving numbers 1 - 99 have 9801 results. Suppose A < B, then A * B = B * A. Then, the multiplication of any two numbers from 1 to 99 actually has only 4901 results. Even so, multiplication is much more complicated than squaring, and the patterns are harder to identify. As we described in the previous chapters, we can list these 4901 answers first, and then maybe we can find the pattern. If you have a solid math foundation, just look at a small set of examples.

If we list the results of the 4901 multiplications in the usual order (from smallest to largest or from largest to smallest), it is hard to see a pattern. Such as

12 x 12 = 144   12 x 13 = 156   12 x 14 = 168   12 x 15 = 180   12 x 16 = 192

and so on.

But, if we list the results of these 4901 multiplications in a different order, a miracle happens. Since we know how to calculate squares quickly, let's start with the squares and use them as an axis to list the results of the multiplications. See the examples below.

9 x 9 = 81	10 x 10 = 100	11 x 11 = 121	12 x 12 = 144
8 x 10 = 80	9 x 11 = 99	10 x 12 = 120	11 x 13 = 143
7 x 11 = 77	8 x 12 = 96	9 x 13 = 117	10 x 14 = 140
6 x 12 = 72	7 x 13 = 91	8 x 14 = 112	9 x 15 = 135
5 x 13 = 63	6 x 14 = 84	7 x 15 = 105	8 x 16 = 128
4 x 14 = 56	5 x 15 = 75	6 x 16 = 96	7 x 17 = 119

3 x 15 = 45	4 x 16 = 64	5 x 17 = 85	6 x 18 = 108
2 x 16 = 32	3 x 17 = 51	4 x 18 = 72	5 x 19 = 95
1 x 17 = 17	2 x 18 = 36	3 x 19 = 57	4 x 20 = 80
	1 x 19 = 19	2 x 20 = 40	3 x 21 = 63
		1 x 21 = 21	2 x 22 = 44
			1 x 23 = 23

Think about it for a while and see if you can see a pattern. In fact, the pattern is not so obvious.

From the following 8 groups of examples, you may be able to see some clues:

99 x 99 = 9801	98 x 98 = 9604	97 x 97 = 9409	96 x 96 = 9216
	97 x 99 = 9603	96 x 98 = 9408	95 x 97 = 9215
		95 x 99 = 9405	94 x 98 = 9212
			93 x 99 = 9207

95 x 95 = 9025	94 x 94 = 8836	93 x 93 = 8649	92 x 92 = 8464
94 x 96 = 9024	93 x 95 = 8835	92 x 94 = 8648	91 x 93 = 8463
93 x 97 = 9021	92 x 96 = 8832	91 x 95 = 8645	90 x 94 = 8460
92 x 98 = 9016	91 x 97 = 8827	90 x 96 = 8640	89 x 95 = 8455
91 x 99 = 9009	90 x 98 = 8820	89 x 97 = 8633	88 x 96 = 8448
	89 x 99 = 8811	88 x 98 = 8624	87 x 97 = 8439
		87 x 99 = 8613	86 x 98 = 8428
			85 x 99 = 8415

If you list all the multiplications related to 91 x 91 in this order (from 90 x 92, 89 x 93 to 83 x 99), all the multiplications related to 90 x 90, and all the multiplications related to 50 x 50, the pattern emerges. This is how I stumbled upon this multiplication pattern.

From the above examples, we can see some clues. Taking the square of 95 as an example, let's rewrite the above formula:
Since $0 = (95 - 95)/2$, $95 \times 95 = 9025 = 9025 - 0 = 95^2 - 0^2 = (95 + 0)^2 - 0^2$
Since $1 = (96 - 94)/2$, $94 \times 96 = 9024 = 9025 - 1 = 95^2 - 1^2 = (94 + 1)^2 - 1^2$
Since $2 = (97 - 93)/2$, $93 \times 97 = 9021 = 9025 - 4 = 95^2 - 2^2 = (93 + 2)^2 - 2^2$
Since $3 = (98 - 92)/2$, $92 \times 98 = 9016 = 9025 - 9 = 95^2 - 3^2 = (92 + 3)^2 - 3^2$
Since $4 = (99 - 91)/2$, $91 \times 99 = 9009 = 9025 - 16 = 95^2 - 4^2 = (91 + 4)^2 - 4^2$

Let's take another example with the square of 92 as the axis.
Since $0 = (92 - 92)/2$, $92 \times 92 = 8464 = 8464 - 0 = 92^2 - 0^2 = (92 + 0)^2 - 0^2$
Since $1 = (93 - 91)/2$, $91 \times 93 = 8463 = 8464 - 1 = 92^2 - 1^2 = (91 + 1)^2 - 1^2$
Since $2 = (94 - 90)/2$, $90 \times 94 = 8460 = 8464 - 4 = 92^2 - 2^2 = (90 + 2)^2 - 2^2$
Since $3 = (95 - 89)/2$, $89 \times 95 = 8455 = 8464 - 9 = 92^2 - 3^2 = (89 + 3)^2 - 3^2$
Since $4 = (96 - 88)/2$, $88 \times 96 = 8448 = 8464 - 16 = 92^2 - 4^2 = (88 + 4)^2 - 4^2$
Since $5 = (97 - 87)/2$, $87 \times 97 = 8439 = 8464 - 25 = 92^2 - 5^2 = (87 + 5)^2 - 5^2$
Since $6 = (98 - 86)/2$, $86 \times 98 = 8428 = 8464 - 36 = 92^2 - 6^2 = (86 + 6)^2 - 6^2$
Since $7 = (99 - 85)/2$, $85 \times 99 = 8415 = 8464 - 49 = 92^2 - 7^2 = (85 + 7)^2 - 7^2$

From the above two examples, we can summarize as follows:
Assuming $A_1 < A_2$, $m = (A_2 - A_1)/2$, then

Formula 6.1: $A_1 * A_2 = (A1 + m)^2 - m^2$

That's how simple the multiplication formula is. Now, we verify three sets of examples using **Formula 6.1**.

First group:
$A_1 = 93$, $A_2 = 93$. Since, $m = (93 - 93) / 2 = 0$, so $93 \times 93 = (93 + 0)^2 - 0^2 = 8649$
$A_1 = 92$, $A_2 = 94$. Since, $m = (94 - 92) / 2 = 1$, so $94 \times 92 = (92 + 1)^2 - 1^2 = 93^2 - 1^2 = 8649 - 1 = 8648$
$A_1 = 91$, $A_2 = 95$. Since, $m = (95 - 91) / 2 = 2$, so $91 \times 95 = (91 + 2)^2 - 2^2 = 93^2 - 2^2 = 8649 - 4 = 8645$
$A_1 = 90$, $A_2 = 96$. Since, $m = (96 - 90) / 2 = 3$, so $90 \times 96 = (90 + 3)^2 - 3^2 = 93^2 - 3^2 = 8649 - 9 = 8640$
$A_1 = 89$, $A_2 = 97$. Since, $m = (97 - 89) / 2 = 4$, so $89 \times 97 = (89 + 4)^2 - 4^2 = 93^2 - 4^2 = 8649 - 16 = 8633$
$A_1 = 88$, $A_2 = 98$. Since, $m = (98 - 88) / 2 = 5$, so $88 \times 98 = (88 + 5)^2 - 5^2 = 93^2 - 5^2 = 8649 - 25 = 8624$
$A_1 = 87$, $A_2 = 99$. Since, $m = (99 - 87) / 2 = 6$, so $89 \times 97 = (89 + 6)^2 - 6^2 = 93^2 - 6^2 = 8649 - 36 = 8613$

Second Group:
Since, m = (90 − 90) / 2 = 0, so 90 x 90 = (90 + 0)² − 0² = 8100
Since, m = (91 − 89) / 2 = 1, so 89 x 91 = (89 + 1)² − 1² = 90² − 1² = 8100 − 1 = 8099
Since, m = (92 − 88) / 2 = 2, so 88 x 92 = (88 + 2)² − 2² = 90² − 2² = 8100 − 4 = 8096
Since, m = (93 − 87) / 2 = 3, so 87 x 93 = (87 + 3)² − 3² = 90² − 3² = 8100 − 9 = 8091
Since, m = (94 − 86) / 2 = 4, so 86 x 94 = (86 + 4)² − 4² = 90² − 4² = 8100 − 16 = 8084
Since, m = (95 − 85) / 2 = 5, so 85 x 95 = (85 + 5)² − 5² = 90² − 5² = 8100 − 25 = 8075
Since, m = (96 − 84) / 2 = 6, so 84 x 96 = (84 + 6)² − 6² = 90² − 6² = 8100 − 36 = 8064
Since, m = (97 − 83) / 2 = 7, so 83 x 97 = (83 + 7)² − 7² = 90² − 7² = 8100 − 49 = 8051
Since, m = (98 − 82) / 2 = 8, so 82 x 98 = (82 + 8)² − 8² = 90² − 8² = 8100 − 64 = 8036
Since, m = (99 − 81) / 2 = 9, so 81 x 99 = (81 + 9)² − 9² = 90² − 9² = 8100 − 81 = 8019

Third group:
Since, m = (12 − 12) / 2 = 0, so 12 x 12 = (12 + 0)² − 0² = 144
Since, m = (13 − 11) / 2 = 1, so 11 x 13 = (11 + 1)² − 1² = 12² − 1² = 144 − 1 = 143
Since, m = (14 − 10) / 2 = 2, so 10 x 14 = (10 + 2)² − 2² = 12² − 2² = 144 − 4 = 140
Since, m = (15 − 9) / 2 = 3, so 9 x 15 = (9 + 3)² − 3² = 12² − 3² = 144 − 9 = 135

Let's first look at how magical the multiplication of magic numbers is.

The product ($A_1 * A_2$) of any two numbers (A_1 and A_2) can be converted into the squared variance of the two numbers. As we observed above:

$$\textbf{Formula 6.1: } A_1 * A_2 = (A_1 + m)^2 - m^2$$

where $A_1 < A_2$ and the magic number **m** = $(A_2 - A_1)/2$. The formula for magic multiplication is so simple! This formula was discovered by me in 2011 through observation. This formula was named by an academician of mathematics in the Chinese Academy of Sciences as "the **Chengqi Magic Multiplication Formula**", or "the **Chengqi Rapid Calculation**" in **abbreviation.**

Even though I found **Formula 6.1** through observation, it is absolutely correct. A formal proof can be found in **Section 8.1**.

Since we are now fluent in mentally squaring 1 to 125, as long as $A_1 + m < 125$, we can easily calculate $(A_1 * A_2)$.

Let's look at an example:
57 x 69 =?
The magic number **m** = (69 − 57)/2 = **6**. So, 57 x 69 = $(57 + 6)^2 - 6^2$ = $63^2 - 6^2$ = 3969 − 36 = 3933

Let's look at another example:
26 x 188 =?
The magic number **m** = (188 - 26)/2 = **81**. So, 26 x 188 = $(26 + 81)^2 - 81^2$ = $107^2 - 81^2$ = 11449 − 6561 = 4888

In the above examples, the numbers need to be either both odd or both even for their magic number to be an integer. If one of the two numbers is odd and the other is even, their magic numbers are not integers. Formula 6.1 is still valid but computing the square of the magic number is more complicated. We can solve it in the following way:

When the mean of two numbers is not an integer (one multiplier is odd and one multiplier is even), their magic number m is not an integer. In other words, in order for the mean of two numbers to be an integer, both numbers must be odd or both numbers must be even. Thus, when one of the two multipliers is odd and the other is even, we can convert the multiplication formula (just do one more addition). For example:

36 x 49 = (35 + 1) x 49 = 35 x 49 + 49 or 36 x 49 = 36 x (48 + 1) = 36 x 48 + 36

Apply the **Formula 6.1** in this chapter to calculate 35 x 49 or 36 x 48. Do one more addition and the problem will be solved.

36 x 49 = 35 x 49 + 49 since the magic number m = (49 − 35)/2 = 14/2 = 7 for 35 x 49,
So 35 x 49 = $(35 + 7)^2 - 7^2$ = $42^2 - 7^2$ = 1764 - 49 = 1715,
So 36 x 49 = 35 x 49 + 49 = 1715 + 49 = 1764
or 36 x 49 = 36 x 48+ 36, because the magic number m = (48 − 36)/2 = 12/2 = 6 for 36 x 48,
So 36 x 48 = $(36 + 6)^2 - 6^2$ = $42^2 - 6^2$ = 1764 - 36 = 1728,
So 36 x 49 = 36 x 48+ 36 = 1728 + 36 = 1764

Can you think of any other way to extend this pattern? Find the limitations of this algorithm, and then find ways to extend it to break its limitations. This is the most important scientific research method.

We abstract the rapid calculation Formula 6.1 for calculating $A_1 * A_2$ into the following four-step calculation:

Step 1: Calculate the magic number $m = (A_2 - A_1)/2$
Step 2: Calculate $(A_1 + m)$ and its squared value;
Step 3: Calculate the square value of the magic number m.
Step 4: Calculate a subtraction: Subtract the result of step 2 from the result of step 3.

Exercise 6.1: In the following exercises, please calculate the magic number of each question first, and then calculate the answer.

46 x 54 =	38 x 42 =	72 x 88 =	94 x 106 =
91 x 109 =	57 x 59 =	56 x 60 =	54 x 62 =
53 x 63 =	51 x 65 =	46 x 58 =	79 x 81 =
62 x 82 =	39 x 99 =	28 x 68 =	37 x 57 =
64 x 84 =	31 x 91 =	68 x 88 =	32 x 52 =
63 x 82 =	40 x 99 =	29 x 68 =	38 x 57 =
65 x 84 =	32 x 91 =	69 x 88 =	33 x 52 =

Section 6.2: Magic Multiplication of Two Numbers When They Differ By < 20 (difficulty 4 *)

Formula 6.1: $A_1 \times A_2 = (A_1 + m)^2 - m^2$

Because the magic number $m = (A_2 - A_1)/2$ and $(A_2 - A_1) < 20$, $m < 10$. The second term of Formula 6.1 is less than 10, and m squared is also less than 100, which is convenient for calculation and subtraction. When the ones digits of two numbers are both odd or both even, the magic number m is a positive integer. Now let's look at the example.

Example: 67 x 83 =?
The magic number m = (83 - 67) / 2 = 16/2 = 8,
So, 67 x 83 = $(67 + 8)^2 - 8^2 = 75^2 - 8^2$ = 5625 – 64 = 5525 + 1 + 99 – 64 = 5526 + 35 = 5561

Example: 57 x 69 =?
The magic number m = (69 – 57) / 2 = 12/2 = 6,
So, 57 x 69 = $(57 + 6)^2 - 6^2 = 63^2 - 6^2$ = 3969 – 36 = 3933

Example: 63 x 67 =?
The magic number m = (67 - 63) / 2 = 4/2 = 2,
So, 63 x 67 = (63 + 2)² – 2² = 65² – 2² = 4225 – 4 = 4221

Example: 38 x 52 =?
The magic number m = (52 - 38) / 2 = 14/2 = 7,
So, 38 x 52 = (38 + 7)² – 7² = 45² – 7² = 2025 – 49 = 1925 + 1 + 99 – 49 = 1926 + 50 = 1976

Example: 36 x 48 =?
The magic number m = (48 - 36) / 2 = 12/2 = 6,
So, 36 x 48 = (36 + 6)² – 6² = 42² – 6² = 1764 – 36 = 1728

Example: 19 x 33 =?
The magic number m = (33 - 19) / 2 = 14/2 = 7,
So, 19 x 33 = (19 + 7)² – 7² = 26² – 7² = 676 – 49 = 627

Example: 14 x 26 =?
The magic number m = (26 - 14) / 2 = 12/2 = 6,
So, 14 x 26 = (14 + 6)² – 6² = 20² – 6² = 400 – 36 = 364

Example: 12 x 14 =?
The magic number m = (14 - 12) / 2 = 1,
So, 12 x 14 = (12 + 1)² – 1² = 13² – 1² = 169 – 1 = 168

Exercise 6.2: In the following exercises, please calculate the magic number of each question first, and then calculate the answer.

76 x 84 =	52 x 66 =	11 x 25 =	13 x 27 =
65 x 73 =	34 x 48 =	66 x 82 =	51 x 69 =

Section 6.3: Magic Multiplication of Two Numbers with Both Odd or Even Tens Digit and the Same Number for the Ones Digit

When two numbers have the same ones digit and their tens digits are both odd or both even, the magic number m = $(A_2 - A_1)/2$ is a multiple of 10. The square of m is followed by a multiple of 100, which is convenient for calculating subtraction. Now take a look at the examples below.

Example: 43 x 83 =?
The magic number m = (83 - 43) / 2 = 40/2 = 20,
So, 43 x 83 = (43 + 20)² – 20² = 63² – 20² = 3969 – 400 = 3569

Example: 17 x 77 =?
The magic number m = (77 - 17) / 2 = 60/2 = 30,
So, 17 x 77 = (17 + 30)² – 30² = 47² – 30² = 2209 – 900 = 1309

Example: 28 x 68 =?
The magic number m = (68 - 28) / 2 = 40/2 = 20,
So, 28 x 68 = (28 + 20)² – 20² = 48² – 20² = 2304 – 400 = 1904

Example: 12 x 52 =?
The magic number m = (52 - 12) / 2 = 40/2 = 20,
So, 12 x 52 = (12 + 20)² – 20² = 32² – 20² = 1024 – 400 = 624

Example: 38 x 98 =?
The magic number m = (98 - 38) / 2 = 60/2 = 30,
So, 38 x 98 = (38 + 30)² – 30² = 68² – 30² = 4624 – 900 = 3724

Example: 13 x 33 =?
The magic number m = (33 - 13) / 2 = 20/2 = 10,
So, 13 x 33 = (13 + 10)² – 10² = 23² – 10² = 529 – 100 = 429

Example: 26 x 86 =?
The magic number m = (86 – 26) / 2 = 60/2 = 30,
So, 26 x 86 = (26 + 30)² – 30² = 56² – 30² = 3136 – 900 = 2236

Example: 14 x 74 =?
The magic number m = (74 - 14) / 2 = 60/2 = 30,
So, 14 x 74 = (14 + 30)² – 30² = 44² – 30² = 1936 – 900 = 1036

Exercise 6.3: In the following exercises, please calculate the magic number of each question first, and then calculate the answer.

24 x 84 =   52 x 92 =   15 x 95 =   13 x 73 =

65 x 85 =   34 x 94 =   36 x 76 =   51 x 71 =

CHAPTER 7: MULTIPLICATION OF SPECIAL NUMBERS (SPECIAL FORMULAS)

Assumed knowledge: The reader is assumed to be proficient in the first six chapters.

This chapter uses seven sections to introduce seven special cases. They are:

Section 7.1: Multiplication of Two Numbers Where the Tens Digit is the Same and the Sum of the Ones Digits Equals 10 (difficulty 2*)
Section 7.2: Multiplication of Two Numbers Where the Tens Digit Differs by 1 and the Sum of the Ones Digits Equals 10 (difficulty 3*)
Section 7.3: Multiplication of Two Numbers Where Sum of the Tens Digits Equals 10 and the Ones Digits are the same (difficulty 2*)
Section 7.4: Multiplication of Any Two Numbers Between 11 and 19 (difficulty 4*)
Section 7.5: Multiplication of Any Two Numbers Between 101 and 109 (difficulty 3*)
Section 7.6: Multiplication of Any Two Numbers Between 91 and 99 (difficulty 4*)
Section 7.7: Multiplication of Any Two Numbers Between 81 and 119 (optional) (difficulty 5*)

Section 7.1: Multiplication of Two Numbers Where the Tens Digit is the Same and the Sum of the Ones Digits Equals 10 (difficulty 2*)

Before explaining this section, please quickly calculate the following results. If you can do it, skip this section and go directly to **Section 7.2**.

51 x 59 =	53 x 57 =	55 x 55 =	54 x 56 =	52 x 58 =
61 x 69 =	63 x 67 =	65 x 65 =	64 x 66 =	62 x 68 =
81 x 89 =	83 x 87 =	85 x 85 =	84 x 86 =	82 x 88 =
41 x 49 =	43 x 47 =	45 x 45 =	44 x 46 =	42 x 48 =
21 x 29 =	23 x 27 =	25 x 25 =	24 x 26 =	22 x 28 =

11 x 19 =	13 x 17 =	15 x 15 =	14 x 16 =	12 x 18 =
31 x 39 =	33 x 37 =	35 x 35 =	34 x 36 =	32 x 38 =
71 x 79 =	73 x 77 =	75 x 75 =	74 x 76 =	72 x 78 =
91 x 99 =	93 x 97 =	95 x 95 =	94 x 96 =	92 x 98 =

Maybe you can mentally work out the results, but it may not be fast enough. I'm sure you'll be able to calculate them quickly after reading this section, perhaps surprising even yourself.

Please see the examples below:
31 x 39 = 1209
32 x 38 = 1216
33 x 37 = 1221
34 x 36 = 1224
35 x 35 = 1225

Note that because 36 x 34 = 34 x 36, 37 x 33 = 33 x 37, ... the algorithm is the same, so we only need to list the first five, the same below.

Do you notice the pattern? If you don't, see the following variants of the equation:
31 x 39 = **3** x (**3** +1) x 100 + 1 x 9 = 12 09
32 x 38 = **3** x (**3** +1) x 100 + 2 x 8 = 12 16
33 x 37 = **3** x (**3** +1) x 100 + 3 x 7 = 12 21
34 x 36 = **3** x (**3** +1) x 100 + 4 x 6 = 12 24
35 x 35 = **3** x (**3** +1) x 100 + 5 x 5 = 12 25

Do you see the pattern now? You should note that the multiplication of two two-digit numbers results in three or four digits.

The pattern is as follows: "if the tens digits are the same, and the sum of the ones digits is equal to 10, the first two digits (or the previous digit) of their product are the product of the tens digit and the number larger than the tens digit by 1. The last two digits are the product of the two ones digits.

In this section, there are three magic numbers, one tens digit and two ones digits. We define the ones digit of the first number as m_1 (magic number 1) and the ones digit of the second number as m_2 (magic number 2), where $m_1 + m_2 = 10$. Define the tens digit as Y (magic number 3). A_1 represents the first number, A_2 represents the second number, Y represents the tens digit, m_1 represents the ones digit of the first number, m_2 represents the ones digit of the second number. So, $A_1 = 10 * Y + m_1$, $A_2 = 10 * Y + m_2$, $m_1 + m_2 = 10$,

Formula 7.1: $A_1 * A_2 = (10 * Y + m_1) * (10 * Y + m_2)$
$= Y * (Y + 1) \times 100 + m_1 * m_2$

Below we use **Formula 7.1** to give a few examples:
Example: **34** x **36** =?
In this example, Y = 3, m_1 = 4, m_2 = 6, so
34 x **36** = Y * (Y + 1) * 100 + m_1 * m_2 = **3** x (**3** + 1) x 100 + 4 x 6 = 1200 + 24 = 12 24

Example: **53** x **57** =?
In this example, Y = 5, m_1 = 3, m_2 = 7, so
53 x **57** = Y * (Y + 1) * 100 + m_1 * m_2 = **5** x (**5** + 1) x 100 + 3 x 7 = 3000 + 21 = 3021

Example: **82** x **88** =?
In this example, Y = 8, m_1 = 2, m_2 = 8, so
82 x **88** = Y * (Y + 1) * 100 + m_1 * m_2 = **8** x (**8** + 1) x 100 + 2 x 8 = 7200 + 16 = 7216

Example: **65** x **65** =?
In this example, Y = 6, m_1 = 5, m_2 = 5, so
65 x **65** = Y * (Y + 1) * 100 + m_1 * m_2 = **6** x (**6** + 1) x 100 + 5 x 5 = 4200 + 25 = 4225

See **Section 8.2** for a detailed proof of **Formula 7.1**.

In **Formula 7.1,** after finding the 3 magic numbers, there are only two steps of calculation. The first step calculates the tens digit multiplied by (the tens digit + 1) and puts it in the first two digits (the magic number 3 is used). The second step calculates the two ones digits (the magic number 1 and the magic number 2 are used) in the last two digits. Therefore, the rapid calculation **Formula 7.1** simplifies the traditional 8 steps calculation of two-digit multiplication to 2 steps of calculation.

Note that in Section 1.2, we have introduced the algorithm for "squaring numbers whose ones digits is 5," calculated using **Formula 1.1.**
e.g., 65 x 65 = 6 x (6 + 1) x 100 + 25 = 42 25.

You will note that "the square of a number whose ones digit is 5" also satisfies the above conditions. That is: the tens digit is the same, and the addition of the ones digit equals 10. So, we can also use the Formula 1.1 for the calculation. In fact, in the **Formula 7.1,** when m_1 = m_2 = 5, the Formula 7.1 degenerates into the **Formula 1.1**. That is, the **Formula 1.1** is a special case of the **Formula 7.1**.

Exercise 7.1: In the following exercises, please find the three magic numbers for each question, and then calculate the answer.

51 x 59 =	53 x 57 =	55 x 55 =	54 x 56 =	52 x 58 =
81 x 89 =	83 x 87 =	85 x 85 =	84 x 86 =	82 x 88 =
21 x 29 =	23 x 27 =	25 x 25 =	24 x 26 =	22 x 28 =
31 x 39 =	33 x 37 =	35 x 35 =	34 x 36 =	32 x 38 =
91 x 99 =	93 x 97 =	95 x 95 =	94 x 96 =	92 x 98 =

Section 7.2: Multiplication of Two Numbers Where the Tens Digit Differs by 1 and the Sum of the Ones Digits Equals 10 (difficulty 3*)

Before reading the explanation of this section, try to quickly calculate the following results. If you can do it, you can skip this section and go directly to **Section 7.3**.

51 x 69 =	53 x 67 =	55 x 65 =	54 x 66 =	52 x 68 =
61 x 79 =	63 x 77 =	65 x 75 =	64 x 76 =	62 x 78 =
81 x 99 =	83 x 97 =	85 x 95 =	84 x 96 =	82 x 98 =
41 x 59 =	43 x 57 =	45 x 55 =	44 x 56 =	42 x 58 =
21 x 39 =	23 x 37 =	25 x 35 =	24 x 36 =	22 x 38 =
11 x 29 =	13 x 27 =	15 x 25 =	14 x 26 =	12 x 28 =
31 x 49 =	33 x 47 =	35 x 45 =	34 x 46 =	32 x 48 =
71 x 89 =	73 x 87 =	75 x 85 =	74 x 86 =	72 x 88 =
91 x 109 =	93 x 107 =	95 x 105 =	94 x 106 =	92 x 108 =

56 x 64 =	57 x 63 =	58 x 62 =	59 x 61 =
66 x 74 =	67 x 73 =	68 x 72 =	69 x 71 =
86 x 94 =	87 x 93 =	88 x 92 =	89 x 91 =
46 x 54 =	47 x 53 =	48 x 52 =	49 x 51 =
26 x 34 =	27 x 33 =	28 x 32 =	29 x 31 =

16 x 24 =   17 x 23 =   18 x 22 =   19 x 21 =

36 x 44 =   37 x 43 =   38 x 42 =   39 x 41 =

76 x 84 =   77 x 83 =   78 x 82 =   79 x 81 =

96 x 104 =   97 x 103 =   98 x 102 =   99 x 101 =

If your calculations are not fast enough, read on.
Let's start with some simple examples:
    60 x 60 = 3600
    59 x 61 = 3599
    58 x 62 = 3596
    57 x 63 = 3591
    56 x 64 = 3584

Do you notice the pattern? Here, I convert them into the following set of expressions:
    59 x **61** = 6^2 x 100 - **1^2** = 3600 - 1 = 3599
    58 x **62** = 6^2 x 100 - **2^2** = 3600 - 4 = 3596
    57 x **63** = 6^2 x 100 - **3^2** = 3600 - 9 = 3591
    56 x **64** = 6^2 x 100 - **4^2** = 3600 - 16 = 3584

This set of patterns is surprisingly simple, but hard to spot. In fact, we only need to look at the larger of the two numbers. That is, the product of the two numbers is equal to the square of the tens digit of the larger number multiplied by 100 minus the square of the ones digit. For example, 58 x 62, the larger number is 62, the tens digit is 6, and the ones digit is 2, then 58 x 62 = 6^2 x 100 – 2^2 = 3600 – 4 = 3594. This pattern can be formally expressed as:

Assuming that the larger number is A_2, its tens digit is Y (magic number 1) and its ones digit is m (magic number 2), then A_2 = (Y * 10) + m. Assuming the smaller number is A_1, its tens digit is (Y - 1) and the ones digit is (10 - m), then A_1 = (Y - 1) x 10 + 10 - m = Y * 10 – 10 + 10 – m = Y * 10 - m. So, Formula 7.2 is:

Formula 7.2: A_1 * A_2 = (Y * 10 – m) * (Y * 10 + m) = Y^2 * 100 – m^2

A detailed proof of **Formula 7.2** is given in **Section 8.3**.
Below, we use **Formula 7.2** to give a few examples:
Example: 23 x 37 =?
In this example, because A_2 = 37, so Y = 3, m = 7,
using **Formula 7.2**:
23 x 37 = Y^2 * 100 – m^2 = 3^2 x 100 - 7^2 = 900 - 49 = 851

Example: 45 x 55 =?
In this example, because A_2 = 55, so Y = 5, m = 5,
using **Formula 7.2**:
45 x 55 = Y^2 * 100 – m^2 = 5^2 x 100 - **5^2** = 2500 - 25 = 2475

Example: 98 x 102 =?
In this example, because A_2 = 102, so Y = 10, m = 2,
using **Formula 7.2**:
98 x 102 = Y^2 * 100 – m^2 = 10^2 x 100 - **2^2** = 10000 - 4 = 9996

Example: 52 x 68 =?
In this example, because A_2 = 68, so Y = 6, m = 8,
using **Formula 7.2**:
52 x 68 = Y^2 * 100 – m^2 = 6^2 x 100 - **8^2** = 3600 - 64 = 3536

In **Formula 7.2**, after finding 2 magic numbers, there are only three steps of calculation. The first step is to calculate the square of the magic number 1, followed by two 0s. The second step is to calculate the square of the magic number 2. The third step is to do a subtraction. So, the rapid calculation Formula 7.2 simplifies the traditional two-digit multiplication to 3 steps of calculation.

The following paragraph is an optional reading for those who do not like subtraction. When we encounter a subtraction borrowing, we can easily convert the subtraction operation to an addition operation, and the principle is as follows:

If the subtrahend is a one-digit number, decompose the minuend into (minuend – 10) + 1 + 9;

If the subtrahend is a two-digit number, decompose the minuend into (minuend – 100) + 1 + 99;

If the subtrahend is a three-digit number, decompose the minuend into (minuend – 1000) + 1 + 999.

Here, we illustrate with a few examples.
3600 − 64 = 3500 + 1 + 99 − 64 = 3501 + 35 = 3536
900 − 49 = 800 + 1 + 99 − 49 = 801 + 50 = 851
1736 − 81 = 1636 + 1 + 99 − 81 = 1637 + 18 = 1655
2809 − 16 = 2709 + 1 + 99 − 16 = 2710 + 83 = 2793
2681 − 4 = 2671 + 1 + 9 − 4 = 2672 + 5 = 2677
2681 − 786 = 1681 + 1 + 999 − 786 = 1682 + 213 = 1895

The above method is useful and will ease the difficulty of borrowing subtraction.

Exercise 7.2: In the following exercises, please find the two magic numbers for each question, and then calculate the answer.

51 x 69 =	53 x 67 =	55 x 65 =	54 x 66 =	52 x 68 =
61 x 79 =	63 x 77 =	65 x 75 =	64 x 76 =	62 x 78 =
81 x 99 =	83 x 97 =	85 x 95 =	84 x 96 =	82 x 98 =
41 x 59 =	43 x 57 =	45 x 55 =	44 x 56 =	42 x 58 =
21 x 39 =	23 x 37 =	25 x 35 =	24 x 36 =	22 x 38 =
11 x 29 =	13 x 27 =	15 x 25 =	14 x 26 =	12 x 28 =
31 x 49 =	33 x 47 =	35 x 45 =	34 x 46 =	32 x 48 =
71 x 89 =	73 x 87 =	75 x 85 =	74 x 86 =	72 x 88 =
91 x 109 =	93 x 107 =	95 x 105 =	94 x 106 =	92 x 108 =

Section 7.3: Multiplication of Two Numbers Where Sum of the Tens Digits Equals 10 and the Ones Digits are the same (difficulty 2*)

Before explaining this section, please try to quickly calculate the following results. If you can do it, you can skip this section and go directly to **Section 7.4**.

15 x 95 =	35 x 75 =	55 x 55 =	45 x 65 =	25 x 85 =
16 x 96 =	36 x 76 =	56 x 56 =	46 x 66 =	26 x 86 =
18 x 98 =	38 x 78 =	58 x 58 =	48 x 68 =	28 x 88 =
14 x 94 =	34 x 74 =	54 x 54 =	44 x 64 =	24 x 84 =
12 x 92 =	32 x 72 =	52 x 52 =	42 x 62 =	22 x 82 =

11 x 91 =	31 x 71 =	51 x 51 =	41 x 61 =	21 x 81 =
13 x 93 =	33 x 73 =	53 x 53 =	43 x 63 =	23 x 83 =
17 x 97 =	37 x 77 =	57 x 57 =	47 x 67 =	27 x 87 =
19 x 99 =	39 x 79 =	59 x 59 =	49 x 69 =	29 x 89 =

Maybe you can mentally work out the result, but it is not fast enough. However, I'm sure you'll be able to calculate them quickly after reading this section.

Please see the examples below:
13 x 93 = 12 09
23 x 83 = 19 09
33 x 73 = 24 09
43 x 63 = 27 09
53 x 53 = 28 09

Similar to the previous section, because 63 x 43 = 43 x 63 and 73 x 33 = 33 x 73 and so on, we will not list out both.

The pattern of the latter two are obvious, and the first two are a bit more difficult. Think about it: what do the first two have to do with the product of two ten-digit numbers?

Did you work out the pattern? There are three magic numbers: two different tens digits and one identical ones digit. The first two digits of the product are directly related to the three magic numbers, and the last two digits are related only to the ones digits. The answer lies in the transformed form of the example below:

13 x 93 = (1 x 9 + **3**) x 100+ **3**^{2} = 12 09
23 x 83 = (2 x 8 + **3**) x 100+ **3**^{2} = 19 09
33 x 73 = (3 x 7 + **3**) x 100+ **3**^{2} = 24 09
43 x 63 = (4 x 6 + **3**) x 100+ **3**^{2} = 27 09
53 x 53 = (5 x 5 + **3**) x 100+ **3**^{2} = 28 09

I hope you have found the pattern. If you haven't found it yet, here is my summary:

The pattern is as follows: when we calculate the multiplication of two numbers where the sum of their tens digits is equal to 10, and two numbers have the same ones digit, the first two digits of the result are the product of the two tens digits plus the ones digit, and the last two digits are the square of the ones digit.

In this section, there are also three magic numbers, which are the two different tens digits and the same ones digit. This pattern only applies to the product of two numbers whose tens digits sum to 10 and whose ones digits are the same. We can formally express it as follows:

We use A_1 to represent the first number, whose tens digit is Y_1 (magic number 1), and A_2 to represent the second number, whose tens digit is Y_2 (magic number 2), the single digit is m (magic number 3). So, $A_1 = 10 * Y_1 + m$, and $A_2 = 10 * Y_2 + m$. when $Y_1 + Y_2 = 10$,

Formula 7.3: $A_1 * A_2 = (10 * Y_1 + m) * (10 * Y_2 + m) = (Y_1 * Y_2 + m) * 100 + m^2$

A rigorous proof of Formula 7.3 can be found in **Section 8.4**.
We use Formula 7.3 to give a few examples:
Example: 23 x 83 =?
In this example, $Y_1 = 2$, $Y_2 = 8$, m = 3, so
23 x 83 = $(Y_1 * Y_2 + m) * 100 + m^2$ = (2 x 8 + 3) x 100 + 3^2 = 1900 + 9 = 1909

Example: 35 x 75 =?
In this example, $Y_1 = 3$, $Y_2 = 7$, m = 5, so
35 x 75 = $(Y_1 * Y_2 + m) * 100 + m^2$ = (3 x 7 + 5) x 100 + 5^2 = 2600 + 25
= 2625

Example: 48 x 68 =?
In this example, $Y_1 = 4$, $Y_2 = 6$, m = 8, so
48 x 68 = $(Y_1 * Y_2 + m) * 100 + m^2$ = (4 x 6 + 8) x 100 + 8^2 = 3200 + 64
= 3264

Example: 56 x 56 =?
In this example, $Y_1 = 5$, $Y_2 = 5$, m = 6, so
56 x 56 = $(Y_1 * Y_2 + m) * 100 + m^2$ = (5 x 5 + 6) x 100 + 6^2 = 3100 + 36
= 3136

In **Formula 7.3**, after finding the 3 magic numbers, there are only three steps of calculation. The first step is to calculate the product of two tens digit numbers (magic number 1 and magic number 2 are used). The second step is to calculate the product plus the ones digit number (magic number 3), and put it in the first two digits. In the third step, calculate the square of the ones digit (magic number 3) and place it in the last two digits. Therefore, the rapid calculation **Formula 7.3** simplifies the traditional 8-step calculation of two-digit multiplication to a 3-step calculation.

Note that the last example in each set of examples involves the squares of 50 - 59. We have introduced the Formula 2.1 in Section 2.1, e.g., 56^2 = (25 + 6) x 100 + 6^2 = 3100 + 36 = 3136. If the Formula 7.3 is used, then 56^2 = (5 x 5 + 6) x 100 + 6^2 = 3100 + 36 = 3136.

Exercise 7.3: In the following exercises, please find the three magic numbers for each question, and then calculate the answer.

15 x 95 =	35 x 75 =	55 x 55 =	4 5 x 65 =	25 x 85 =
18 x 98 =	38 x 78 =	58 x 58 =	48 x 68 =	28 x 88 =
12 x 92 =	32 x 72 =	52 x 52 =	42 x 62 =	22 x 82 =
13 x 93 =	33 x 73 =	53 x 53 =	43 x 63 =	23 x 83 =
19 x 99 =	39 x 79 =	59 x 59 =	49 x 69 =	29 x 89 =

Section 7.4: Multiplication of Any Two Numbers Between 11 and 19 (difficulty 4*)

Before explaining this section, please try to quickly calculate the following results. If you can do it, you can skip this section and go directly to **Section 7.5**.

11 x 15 =	12 x 15 =	13 x 15 =	14 x 15 =	15 x 15 =
11 x 16 =	12 x 16 =	13 x 16 =	14 x 16 =	15 x 16 =
11 x 18 =	12 x 18 =	13 x 18 =	14 x 18 =	15 x 18 =
11 x 14 =	12 x 14 =	13 x 14 =	14 x 14 =	11 x 11 =
11 x 12 =	12 x 12 =	11 x 13 =	12 x 13 =	16 x 16 =
11 x 17 =	12 x 17 =	13 x 17 =	14 x 17 =	15 x 17 =
11 x 19 =	12 x 19 =	13 x 19 =	14 x 19 =	15 x 19 =
16 x 17 =	17 x 17 =	13 x 13 =	16 x 1 8 =	17 x 18 =
16 x 19 =	17 x 19 =	18 x 18 =	18 x 19 =	19 x 19 =

If your calculations are not fast enough, read on.
Please see the examples below:
11 x 12 = 132
12 x 14 = 168
13 x 17 = 221
14 x 19 = 266
15 x 16 = 240
16 x 18 = 288

The pattern is not obvious. See the conversion form below:

11 x **12** = (11 + **2**) x 10 + 1 x **2** = 130 + 2 = 132
12 x **14** = (12 + **4**) x 10 + 2 x **4** = 160 + 8 = 168
13 x **17** = (13 + **7**) x 10 + 3 x **7** = 200 + 21 = 221
14 x **19** = (14 + **9**) x 10 + 4 x **9** = 230 + 36 = 266
15 x **16** = (15 + **6**) x 10 + 5 x **6** = 210 + 30 = 240
16 x **18** = (16 + **8**) x 10 + 6 x **8** = 240 + 48 = 288

In the above examples, there are two magic numbers: the two ones digits. The pattern is as follows: the product of the two numbers is the first number plus the magic number (ones digit) of the second number multiplied by 10, plus the product of the two magic numbers (two ones digits). This formula can be expressed formally as follows:

Suppose the first number is A_1, its ones digit (magic number) is m_1, $A_1 = 10 + m_1$. The second number A_2, whose ones digit (magic number) is m_2, $A_2 = 10 + m_2$,

Formula 7.4: $A_1 * A_2 = (10 + m_1) * (10 + m_2) = (A_1 + m_2) \times 10 + m_1 . m_2$

Formula 7.4 can also be re-written as:

Formula 7.4: $A_1 * A_2 = (10 + m_1) * (10 + m_2) = (A_2 + m_1) \times 10 + m_1 . m_2$

A detailed proof of Formula 7.4 is given in Section 8.5.
It can be better understood in the following way:

A_1 m_1
  ✗   → à $A_1 * A_2 = (A_1 + m_2) \times 10 + m_1 . m_2$ or
         → à $A_1 * A_2 = (A_2 + m_1) \times 10 + m_1 . m_2$
A_2 m_2

Her are a few examples using **Formula 7.4**:
Example: 11 x 16 =?
11 1
  ✗   → à 11 x 16 = (11 + 6) x 10 + 1x6 = 17 x 10 + 6 = 176 or
         → à 11 x 16 = (16 + 1) x 10 + 1x6 = 17 x 10 + 6 = 176
16 6

In this example, $A_1 = 11$, $m_1 = 1$, $A_2 = 16$, $m_2 = 6$, using **Formula 7.4**:
11 x 16 = $(A_1 + m_2) \times 10 + m_1 . m_2$ = (11 + 6) x 10 + 1 x 6 = 170 + 6 = 176

Example: 13 x 14 =?
In this example, $A_1 = 13$, $m_1 = 3$, $A_2 = 14$, $m_2 = 4$, using Formula 7.4:
13 x 14 = $(A_1 + m_2)$ x 10 + $m_1 . m_2$ = (13 + 4) x 10 + 3 x 4 = 170 + 12 = 182

Example: 14 x 19 =?
In this example, $A_1 = 14$, $m_1 = 4$, $A_2 = 19$, $m_2 = 9$, using Formula 7.4:
14 x 19 = $(A_1 + m_2)$ x 10 + $m_1 . m_2$ = (14 + 9) x 10 + 4 x 9 = 230 + 36 = 266

Sometimes it is slightly easier to calculate with the deformation formula. For example, 12 x 19 can be calculated in two ways:
1. 12 x 19 = (12 + 9) x 10 + 2 x 9 = 210 + 18 = 228
2. 12 x 19 = (19 + 2) x 10 + 2 x 9 = 210 + 18 = 228

You can choose either way to calculate it.

In **Formula 7.4**, after finding the two ones digits, there are only three steps of calculation. The first step calculates the first number plus the ones digit of the second number, followed by a 0. The second step calculates the product of the two ones digits. The third step is an addition. Therefore, the rapid calculation **Formula 7.4** simplifies the traditional 8 step calculation of two-digit multiplication to 3 steps of calculation.

The content of this section will be used frequently in the future, so please master it proficiently. If you are not familiar with it, please read it several times.

Exercise 7.4: In the following exercises, please find the two magic numbers for each question, and then calculate the answer.

11 x 15 =	12 x 15 =	13 x 15 =	14 x 15 =	15 x 15 =
11 x 16 =	12 x 16 =	13 x 16 =	14 x 16 =	15 x 16 =
11 x 18 =	12 x 18 =	13 x 18 =	14 x 18 =	15 x 18 =
11 x 14 =	12 x 14 =	13 x 14 =	14 x 14 =	16 x 16 =
11 x 17 =	12 x 17 =	13 x 17 =	14 x 17 =	15 x 17 =
11 x 19 =	12 x 19 =	13 x 19 =	14 x 19 =	15 x 19 =
16 x 17 =	17 x 17 =	16 x 19 =	17 x 19 =	19 x 19 =
16 x 18 =	17 x 18 =	18 x 18 =	18 x 19 =	11 x 11 =

Section 7.5: Multiplication of Any Two Numbers Between 101 and 109 (difficulty 3*)

Before explaining this section, please try to quickly calculate the following results. If you can do it, you can skip this section and go directly to **Section 7.6**.

101 x 105 =	102 x 105 =	103 x 105 =	104 x 105 =	105 x 105 =
101 x 106 =	102 x 106 =	103 x 106 =	104 x 106 =	105 x 106 =
101 x 108 =	102 x 108 =	103 x 108 =	104 x 108 =	105 x 108 =
101 x 104 =	102 x 104 =	103 x 104 =	104 x 104 =	101 x 101 =
101 x 102 =	102 x 102 =	101 x 103 =	102 x 103 =	106 x 106 =
101 x 107 =	102 x 107 =	103 x 107 =	104 x 107 =	105 x 107 =
101 x 109 =	102 x 109 =	103 x 109 =	104 x 109 =	105 x 109 =
106 x 107 =	107 x 107 =	106 x 108 =	107 x 108 =	108 x 108 =
106 x 109 =	107 x 109 =	108 x 109 =	109 x 109 =	

If your calculations are not fast enough, read on.
Please see the examples below:
101 x 104 = 10504
102 x 103 = 10506
103 x 106 = 10918
104 x 107 = 11128
106 x 108 = 11448
107 x 108 = 11556
108 x 109 = 11772
109 x 109 = 11881

Do you note the pattern? What is the magic number? Re-transform the above examples to make it clearer.

101 x 104 = (101 + 4) x 100 + 1 x 4 = 10500 + 4 = 10504
102 x 103 = (102 + 3) x 100 + 2 x 3 = 10500 + 6 = 10506
103 x 106 = (103 + 6) x 100 + 3 x 6 = 10900 + 18 = 10918
104 x 107 = (104 + 7) x 100 + 4 x 7 = 11100 + 28 = 11128
106 x 108 = (106 + 8) x 100 + 6 x 8 = 11400 + 48 = 11448
107 x 108 = (107 + 8) x 100 + 7 x 8 = 11500 + 56 = 11556
108 x 109 = (108 + 9) x 100 + 8 x 9 = 11700 + 72 = 11772
109 x 109 = (109 + 9) x 100 + 9 x 9 = 11800 + 81 = 11881

In the above examples, there are two magic numbers, which are the difference between each of the two numbers and 100, or the two ones digits. Remember these two magic numbers, and the product is simple. The product pattern is: the first three digits of the product are the first number plus the magic number of the second number, and the last two digits are the product of the magic numbers. We can explain the formula as follows:

Suppose the first number A_1 has magic number m_1, where $A_1 = 100 + m_1$, or $m_1 = A_1 - 100$. The second number A_2 has magic number m_2, where $A_2 = 100 + m_2$, or $m_2 = A_2 - 100$

Formula 7.5: $A_1 * A_2 = (100 + m_1) * (100 + m_2) = (A_1 + m_2) \times 100 + m_1 * m_2$

A detailed proof of Formula 7.5 is given in **Section 8.6**.

Formula 7.5 can also be re-written as:

Formula 7.5: $A_1 * A_2 = (100 + m_1) * (100 + m_2) = (A_2 + m_1) \times 100 + m_1 * m_2$

It can be better understood in the following way:

$$A_1 \quad m_1$$
$$A_2 \quad m_2$$

→ $A_1 * A_2 = (A_1 + m_2) \times 100 + m_1 * m_2$ or
→ $A_1 * A_2 = (A_2 + m_1) \times 100 + m_1 * m_2$

Using **Formula 7.5**, let's take a look at a few more examples.

Example: 102 x 104 =?

    102 2

→ $102 \times 104 = (102 + 4) \times 100 + 2 \times 4 = 106 \times 100 + 8 = 10608$

or

→ $102 \times 104 = (104 + 2) \times 100 + 2 \times 4 = 106 \times 100 + 8 = 10608$

    104 4

Because $A_1 = 102$, $m_1 = 2$, $A_2 = 104$, $m_2 = 4$,
$102 \times 104 = (A_1 + m_2) \times 100 + m_1 * m_2 = (102 + 4) \times 100 + 2 \times 4 = 10600 + 8 = 10608$

Example: 105 x 108 =?
Since $A_1 = 105$, $m_1 = 5$, $A_2 = 108$, $m_2 = 8$,
$105 \times 108 = (A_1 + m_2) \times 100 + m_1 * m_2 = (105 + 8) \times 100 + 5 \times 8 = 11300 + 40 = 11340$

Example: 106 x 109 =?
Because $A_1 = 106$, $m_1 = 6$, $A_2 = 109$, $m_2 = 9$, so
$106 \times 109 = (A_1 + m_2) \times 100 + m_1 * m_2 = (106 + 9) \times 100 + 6 \times 9 = 11500 + 54 = 11554$

In **Formula 7.5**, after finding the 2 ones digits, there are only two steps of calculation. The first step is to calculate the first number plus the ones digit of the second number, add two 0s after it, and then put it in the first three digits. In the second step, calculate the product of the two ones digits and put it in the last two digits. Since there is no carrying, the addition in the third step is not needed. Therefore, the rapid calculation Formula 7.5 simplifies the traditional three-digit multiplication to 2 steps of calculation.

Do you notice? If $A_1 = A_2 = A$, $m_1 = m_2 = m$, and **Formula 7.5** can be evolved as

$$A_1 * A_2 = A^2 = (100 + m) * (100 + m) = (A + m) \times 100 + m^2$$

This is the **Formula 2.2** of **Section 2.3**.

$$\text{Formula 2.2: } A^2 = (A + m) \times 100 + m^2$$

In other words, **Formula 2.2** is a special case of **Formula 7.5**.

Since we have learned the multiplication of any two numbers between 11 and 19 in the previous section (Section 7.4), you can try the multiplication of the following numbers. Formula 7.5 still applies, except sometimes the product of two magic numbers is greater than 100, so there may be carrying.

101 x 112 =   103 x 114 =   105 x 115 =   108 x 112 =

111 x 112 =   114 x 115 =   115 x 118 =   118 x 118 =

Here is just one example for reference:
Example: 115 x 118 =?

115   15

→ 115 x 118 = (115 + 18) x 100 + 15 x 18 = 133 x 100 + 270
                    = 13570

or

→ 115 x 118 = (118 + 15) x 100 + 15 x 18 = 133 x 100 + 270
                    = 13570

118   18

If you think it is too difficult, skip this part. It will not affect the subsequent reading.

Exercise 7.5: In the following exercises, please find the two magic numbers for each question, and then calculate the answer.

101 x 105 =   102 x 105 =   103 x 105 =   104 x 105 =   105 x 105 =

101 x 108 =   102 x 108 =   103 x 108 =   104 x 108 =   105 x 108 =

101 x 102 =   102 x 102 =   101 x 103 =   102 x 103 =   106 x 106 =

101 x 109 =   102 x 109 =   103 x 109 =   104 x 109 =   105 x 109 =

106 x 107 =   107 x 107 =   106 x 108 =   107 x 108 =   108 x 108 =

Try the following four questions:

103 x 117 =   114 x 118 =   106 x 119 =   119 x 119 =

Section 7.6: Multiplication of Any Two Numbers Between 91 – 99 (difficulty 4*)

Before explaining this section, please try to quickly calculate the following results. If you can do it, skip this section and go straight to Chapter 8.

91 x 95 =   92 x 95 =   93 x 95 =   94 x 95 =   95 x 95 =

91 x 96 =   92 x 96 =   93 x 96 =   94 x 96 =   95 x 96 =

91 x 98 =   92 x 98 =   93 x 98 =   94 x 98 =   95 x 98 =

91X 94 =   92 x 94 =   93 x 94 =   94 x 94 =   91 x 91 =

91 x 92 =   92 x 92 =   91 x 93 =   92 x 93 =   96 x 96 =

91 x 97 =   92 x 97 =   93 x 97 =   94 x 97 =   95 x 97 =

91 x 99 =   92 x 99 =   93 x 99 =   94 x 99 =   95 x 99 =

96 x 97 =   97 x 97 =   96 x 9 8 =   97 x 98 =   98 x 98 =

96 x 99 =   97 x 99 =   98 x 99 =   99 x 99 =

If your calculations are not fast enough, read on.
Please see the examples below:
91 x 97 = 8827
93 x 93 = 8649
94 x 97 = 9118
95 x 96 = 9120

Did you notice the product pattern? Before moving on, let's review Section 2.4. In Section 2.4, we learned that $93^2 = ?$

$93^2 = (93 + (-7)) \times 100 + (-7)^2 = 8649$

Let's transform the above example to see it more clearly:
91 x 97 = (91 + (-3)) x 100 + (-9) x (-3) = 8800 + 27 = 8827
93 x 93 = (93 + (-7)) x 100 + (-7) x (-7) = 8600 + 49 = 8649
94 x 97 = (94 + (-3)) x 100 + (-6) x (-3) = 9100 + 18 = 9118
95 x 96 = (95 + (-4)) x 100 + (-5) x (-4) = 9100 + 20 = 9120

As you can see from the above examples, the multiplication of any two numbers between 91 – 99 is similar to the squaring of numbers between 91 – 99. The difference is that in this section, two magic numbers are used. Please see if the pattern is as follows:

For the product of the two numbers between 91 - 99, the two magic numbers corresponding to the two numbers are the respective differences between the two numbers and 100. The product pattern is: the first two digits are the first number plus the magic number of the second number (because the magic number is negative, it looks like a subtraction), and the last two digits are the product of the two magic numbers.

Suppose the first number A_1 with its magic number being m_1. $A_1 = 100 + m_1$, or $m_1 = A_1 - 100$. The second number A_2 has the magic number m_2. Then, $A_2 = 100 + m_2$, or $m_2 = A_2 - 100$. Formula 7.5 is still used.

Formula 7.5: $A_1 * A_2 = (100 + m_1) * (100 + m_2) = (A_1 + m_2) \times 100 + m_1 * m_2$

Because A_1 and A_2 are both less than 100 and greater than 90, m_1 and m_2 are both negative and between -1 and -9.

This formula is the one used in the previous section (Section 7.5).
Can it be better understood in the following way:

$A_1 \quad m_1$
$\quad \searrow \nearrow \quad$ → $A_1 * A_2 = (A_1 + m_2) \times 100 + m_1 . m_2$ or
$\quad \nearrow \searrow \quad$ → $A_1 * A_2 = (A_2 + m_1) \times 100 + m_1 . m_2$
$A_2 \quad m_2$

We use Formula 7.5 to give two more examples. Remember that the magic number must be found first, and then you can do the calculations using the magic number.

Example: 91 x 93 =?

91 -9

→ 91 x 93 = (91 + (-7)) x 100 + (-9) x (-7) = 84 x 100 + 63
= 8463

or

→ 91 x 93 = (93 + (-9)) x 100 + (-9) x (-7) = 84 x 100 + 63
= 8463

93 -7

The first number is $A_1 = 91$, $m_1 = 91 - 100 = -9$, (note, not 1)
The second number is $A_2 = 93$, $m_2 = 93 - 100 = -7$, (note, not 3)
So, $91 \times 93 = (A_1 + m_2) \times 100 + m_1 * m_2 = (91 + (-7)) \times 100 + (-9) \times (-7) = 8400 + 63 = 8463$

Example: $92 \times 96 = ?$
The first number is $A_1 = 92$, $m_1 = 92 - 100 = -8$, (note, not 2)
The second number is $A_2 = 96$, $m_2 = 96 - 100 = -4$, (note, not 6)
So, $92 \times 96 = (A_1 + m_2) \times 100 + m_1 * m_2 = (92 + (-4)) \times 100 + (-8) \times (-4) = 8800 + 32 = 8832$

In the above example, the two magic numbers are first calculated in two steps, and then there are only two steps remaining for the calculation. The first step is to calculate the first number plus the magic number of the second number, and then two 0s are added at the back. The result is then placed in the first two digits. In the second step, the product of the two magic numbers is calculated and placed in the last two digits. Since there is no carrying, the addition in the third step is not needed. There are 4 steps in total, so the rapid calculation Formula 7.5 simplifies the traditional two-digit multiplication from 8 steps to 4 steps.

Since we learned how to multiply any two numbers between 11 and 19 in Section 7.4, you can try multiplying the following numbers. Formula 7.5 still applies, except sometimes the product of two magic numbers is greater than 100, so there may be carrying.

$81 \times 92 = $ $83 \times 94 = $

$85 \times 95 = $ $88 \times 92 = $

$81 \times 82 = $ $84 \times 85 = $

$85 \times 88 = $ $88 \times 88 = $

Here is an example demonstration: $83 \times 94 = ?$

83 -17

→ $83 \times 94 = (83 + (-6)) \times 100 + (-17) \times (-6) = 77 \times 100 + 102$
    $= 7802$

or

→ $83 \times 94 = (94 + (-17)) \times 100 + (-17) \times (-6) = 77 \times 100 + 102$
    $= 7802$

94 -6

If you think it is too difficult, skip this part. It will not affect the subsequent reading.

Exercise 7.6: In the following exercises, please find the two magic numbers for each question, and then calculate the answer.

91 x 95 = 92 x 95 = 93 x 95 = 94 x 95 = 95 x 95 =
91 x 98 = 92 x 98 = 93 x 98 = 94 x 98 = 95 x 98 =
91 x 92 = 92 x 92 = 91 x 93 = 92 x 93 = 96 x 96 =
91 x 99 = 92 x 99 = 93 x 99 = 94 x 99 = 95 x 99 =
96 x 97 = 97 x 97 = 96 x 98 = 97 x 98 = 98 x 98 =

Try the following four calculations:

83 x 97 =   84 x 98 =   86 x 89 =   89 x 89 =

Section 7.7: Multiplication of Any Two numbers Between 81 and 119 (optional) (difficulty level 5*)

We have already introduced the multiplication of two numbers above 100 in Section 7.5. Section 7.6 introduced the multiplication of two numbers under 100. In this section, we introduce the multiplication of a number below 100 and a number above 100. Because A * B = B * A, we will not introduce the multiplication of a number above 100 and a number below 100.

Suppose the first number $A_1 < 100$. Its magic number is m_1, and $A_1 = 100 + m_1$, or $m_1 = A_1 - 100$. m_1 is a negative number. The second number $A_2 > 100$. Its magic number is m_2 and $A_2 = 100 + m_2$, or $m_2 = A_2 - 100$. m_2 is a positive number. Formula (7-5) is still valid.

Formula 7.5: $A_1 * A_2 = (100 + m_1) * (100 + m_2) = (A_1 + m_2) \times 100 + m_1 * m_2$

It can be better understood in the following way:

$A_1 \quad m_1$
$\quad\quad\searrow\nearrow \quad \rightarrow A_1 * A_2 = (A_1 + m_2) \times 100 + m_1 . m_2$ or
$\quad\quad\nearrow\searrow \quad \rightarrow A_1 * A_2 = (A_2 + m_1) \times 100 + m_1 . m_2$
$A_2 \quad m_2$

Let's explain it with four examples:
Example: 93 x 104 =?
93   -7
　　　　↘   → 93 x 104 = (93 + 4) x 100 + (-7) x 4 = 97 x 100 + (-28)
　　　　　　　　　　　　= 9672
104   4

Example: 93 x 114 =?
93   -7
　　　　↘   → 93 x 114 = (93 +14) x 100 + (-7) x 14 = 107 x 100 + (-98)
　　　　　　　　　　　　= 10602
114   14

Example: 83 x 104 =?
83   -17
         → 83 x 104 = (83 + 4) x 100 + (-17) x 4 = 87 x 100 + (-68)
                  = 8632
104   4

Example: 83 x 114 =?
83   -17
         → 83 x 114 = (83 + 14) x 100 + (-17) x 14 = 97 x 100 + (-238)
                  = 9462
114   14

Exercise 7.7: In the following exercises, please find the two magic numbers for each question, and then calculate the answer.

91 x 105 =	92 x 105 =	93 x 105 =	94 x 105 =	95 x 105 =
91 x 118 =	92 x 118 =	93 x 118 =	94 x 118 =	95 x 118 =
81 x 102 =	82 x 102 =	83 x 102 =	84 x 102 =	85 x 102 =
81 x 119 =	82 x 119 =	83 x 119 =	84 x 119 =	85 x 119 =

Comprehensive exercises for this chapter:

13 x 17 =	35 x 35 =	73 x 77 =	94 x 96 =
13 x 93 =	25 x 85 =	37 x 77 =	49 x 69 =
19 x 21 =	38 x 42 =	55 x 65 =	74 x 86 =
11 x 14 =	13 x 18 =	16 x 16 =	18 x 19 =
103 x 106 =	108 x 116 =	113 x 116 =	108 x 108 =
91 x 94 =	93 x 98 =	96 x 96 =	98 x 99 =
91 x 108 =	92 x 107 =	93 x 106 =	94 x 105 =
91 x 118 =	92 x 117 =	93 x 116 =	94 x 115 =
81 x 109 =	82 x 108 =	83 x 107 =	84 x 106 =

CHAPTER 8: PROOF OF THE QUICK FORMULA FOR MAGIC MULTIPLICATION (DIFFICULTY 5*)

Section 8.1: Proof of Formula 6.1 for the Universal Multiplication

Suppose A_1 is any number, A_2 is another arbitrary number, where $A_1 < A_2$, $m = (A_2 - A_1)/2$. The universal Formula 6.1 for calculating $A_1 * A_2$ is as the follows (see **Chapter 6**):

$$\text{Formula 6.1: } A_1 * A_2 = (A_1 + m)^2 - m^2$$

The proof is as follows:
The right side of Formula 6.1 is
$(A_1 + m)^2 - m^2$
$= (A_1 + m - m) * (A_1 + m + m)$
$= A_1 * (A_1 + 2m)$
$= A_1 * (A_1 + 2 * (A_2 - A_1)/2)$
$= A_1 * (A_1 + A_2 - A_1)$
$= A_1 * A_2$

So, the right side of Formula 6.1 = the left side of Formula 6.1. Quod erat demonstrandum.

Section 8.2: Proof of Formula 7.1 for the Multiplication of Two Numbers Where the Tens digits are the Same and the Sum of the Ones Digits Equals 10

Assuming that A_1 is an arbitrary two-digit number, the tens digit is represented by Y, and the ones digit is represented by m_1, then $A_1 = 10 * Y + m_1$, where $0 < m_1 < 10$. Assuming that A_2 is another two-digit number, the tens digit is also represented by Y (the tens digits of the two numbers are the same), and the ones digit is represented by m_2, then $A_2 = 10 * Y + m_2$, where $0 < m_2 < 10$, and $m_1 + m_2 = 10$. The quick Formula 7.1 to calculate $A_1 * A_2$ is as follows (see **Section 7.1**):

$$\text{Formula 7.1: } A_1 * A_2 = (10 * Y + m_1) * (10 * Y + m_2) = Y * (Y + 1) * 100 + m_1 * m_2$$

The proof is as follows:
We know: $(a + b) * (c + d) = a * c + a * d + b * c + b * d$
because $m_1 + m_2 = 10$
So $A_1 * A_2 = (10 * Y + m_1) * (10 * Y + m_2)$
$= 10 * Y * 10 * Y + 10 * Y * m_2 + m_1 * 10 * Y + m_1 * m_2$

$= 100 * Y^2 + 10 * Y * (m_1 + m_2) + m_1 * m_2$
$= 100 * Y^2 + 100 * Y + m_1 * m_2$
$= 100 * Y * (Y + 1) + m_1 * m_2$
$= Y * (Y + 1) * 100 + m_1 * m_2$

So, we prove the Formula 7.1: $A_1 * A_2 = Y * (Y + 1) * 100 + m_1 * m_2$

Section 8.3: Proof of Formula 7.2 for the Multiplication of Two Numbers Where the Tens Digit Differs by 1 and the Sum of the Ones Digits Equals 10

Suppose $A_1 < A_2$, A_2 is a larger number, its tens digit is Y, and its digit is m, then $A_2 = (Y * 10) + m$. A_1 is a smaller number, its tens digit is $(Y - 1)$, and its ones digit is $(10 - m)$ (because the addition of the two ones digits equals 10), then $A_1 = (Y - 1) * 10 + 10 - m = Y * 10 - m$. The quick Formula 7.2 to calculate $A_1 * A_2$ is as follows (see Section 7.2):

Formula 7.2: $A_1 * A_2 = (Y * 10 - m) * (Y * 10 + m) = Y^2 * 100 - m^2$

The proof is as follows:
We know: $(a - b) * (a + b) = a^2 - b^2$
So $A_1 * A_2 = (Y * 10 - m) * (Y * 10 + m)$
$= Y^2 * 100 - m^2$
So, we prove Formula 7.2: $A_1 * A_2 = Y^2 * 100 - m^2$

Section 8.4: Proof of Formula 7.3 for the Multiplication of Two Numbers Where Sum of the Tens Digits Equals 10 and the Ones Digits are the same

Assuming that A_1 is any two-digit number, the tens digit is represented by Y_1, and the ones digit is represented by m, then $A_1 = 10 * Y_1 + m$, where $0 < m < 10$. Assuming that A_2 is another two-digit number, the tens digit is represented by Y_2, and $Y_1 + Y_2 = 10$, and the ones digit is represented by m (the ones digits of both numbers are the same), then $A_2 = 10 * Y_2 + m$. The **Formula 7.3** calculates $A_1 * A_2$ is as follows (see **Section 7.3**)

Formula 7.3: $A_1 * A_2 = (10 * Y_1 + m) * (10 * Y_2 + m) = (Y_1 * Y_2 + m) * 100 + m^2$

The proof is as follows:
We know: $(a + b) * (c + d) = a * c + a * d + b * c + b * d$
because $Y_1 + Y_2 = 10$
So $A_1 * A_2 = (10 * Y_1 + m) * (10 * Y_2 + m)$
$= 10 * Y_1 * 10 * Y_2 + 10 * Y_1 * m + m * 10 * Y_2 + m^2$
$= 100 * Y_1 * Y_2 + 10 * (Y_1 + Y_2) * m + m^2$
$= 100 * Y_1 * Y_2 + 100 * m + m^2$
$= (Y_1 * Y_2 + m) * 100 + m^2$

So, we prove the Formula 7.3: $A_1 * A_2 = (Y_1 * Y_2 + m) * 100 + m^2$

Section 8.5: Proof of Formula 7.4 for the Multiplication of Any Two Numbers between 11 and 19

Assuming $11 \leq A1 \leq 19$, A_1 can be represented as $A_1 = 10 + m_1$, where $1 \leq m_1 \leq 9$. Suppose A_2 is another number, $11 \leq A_2 \leq 19$. A_2 can be represented as $A_2 = 10 + m_2$, of which $1 \leq m_2 \leq 9$. The Formula 7.4 to calculate $A_1 * A_2$ is (see **Section 7.4**)

Formula 7.4: $A_1 * A_2 = (10 + m_1) * (10 + m_2) = (A_1 + m_2) \times 10 + m_1 * m_2$

The proof is as follows:
$A_1 * A_2 = (10 + m_1) * (10 + m_2)$
$= (10 + m_1) * 10 + (10 + m_1) * m_2$
$= A_1 * 10 + 10 * m_2 + m_1 * m_2$
$= (A_1 + m_2) * 10 + m_1 * m_2$

So, we prove the Formula 7.4: $A_1 * A_2 = (A_1 + m_2) \times 10 + m_1 * m_2$

Formula 7.4 can also be written as

Formula 7.4: $A_1 * A_2 = (10 + m_1) * (10 + m_2) = (A_2 + m_1) \times 10 + m_1 * m_2$

Can you see the slight difference between the two expressions in the above **Formula 7.4**? Can you justify the second expression?

Section 8.6: Proof of Formula 7.5 for the Multiplication of Two Numbers Around 100

Assume $75 < A_1 < 125$, then A_1 can be expressed as $A_1 = 100 + m_1$, where $-25 < m_1 < 25$. Suppose A_2 is another number and $75 < A_2 < 125$, then A_2 can be expressed as $A_2 = 100 + m_2$, where $-25 < m_2 < 25$. The Formula 7.5 to calculate $A_1 * A_2$ is (see **Sections 7.5, 7.6**, and **7.7**)

Formula 7.5: $A_1 * A_2 = (100 + m_1) * (100 + m_2) = (A_1 + m_2) \times 100 + m_1 * m_2$

The proof is as follows:
$A_1 * A_2 = (100 + m_1) * (100 + m_2)$
$= (100 + m_1) * 100 + (100 + m_1) * m_2$
$= A_1 * 100 + 100 * m_2 + m_1 * m_2$
$= (A_1 + m_2) * 100 + m_1 * m_2$

So, we prove the Formula 7.5: $A_1 * A_2 = (A_1 + m_2) \times 100 + m_1 * m_2$

Formula 7.5 can also be written as

Formula 7.5: $A_1 * A_2 = (100 + m_1) * (100 + m_2) = (A_2 + m_1) \times 100 + m_1 * m_2$

Can you see the slight difference between the two expressions in the above **Formula 7.5?** Can you justify the second expression?

PART III
MAGIC SQUARE ROOT SPEED ALGORITHM

The square root calculation (or called square root finding $\sqrt{\Box}$,) is a difficult calculation for most people especially when it comes to mental arithmetic. In fact, the square root calculation is the inverse calculation of the square. Since we learned the rapid calculation for squaring in Part I of this book, we can use the square approximation method, and the square root calculation becomes simple.

CHAPTER 9: SQUARE ROOTS WITHIN 1,000,000 (DIFFICULTY 5*)

Can you mentally calculate the square root of any number up to 1,000,000? It sounds impossible now, but when you finish this chapter, you will find it is a simple thing to do.

When we learned the square calculation in Part I of this book, the results are all definite results without errors. When we take the square root, the square roots of most numbers are approximate, and the key point here is that we limit the margin of error.

For example: we randomly choose a number $\sqrt{3415} \cong ?$ We can use the calculator to get $\sqrt{3415} \cong 58.44$. We can mentally calculate $\sqrt{3415} \cong$ as follows.

If the error is ≤ 5, $\sqrt{3415} \cong 55$. The error is |58.44 − 55| = 3.44 < 5.
If the error is ≤ 2.5, $\sqrt{3415} \cong 57.5$. The error is |58.44 − 57.5| = 0.94 < 2.5.
If the error is ≤ 0.5, $\sqrt{3415} \cong 58.5$. The error is |58.44 − 58.5| = 0.06 < 0.5.
If the error is ≤ 0.25, $\sqrt{3415} \cong 58.25$. The error is |58.44 − 58.25| = 0.19 < 0.25.

You may be curious to know how the above four results are calculated mentally, and whether it is difficult to learn. The answer is No. Please read **Section 9.1** for the ideas and principles, then read **Section 9.2 to 9.4** for the calculation steps.

The magic number square root rapid calculation is the outstanding contribution of this book. I have never seen any other rapid calculation method that approximates square roots in mental arithmetic. The rapid calculation for finding the square root of magic numbers in this book is simple and easy to learn. The threshold to learn the method is so low that as long as you have elementary school arithmetic, you can learn it. I made this discovery in October 2021.

We divide this chapter into four sections, step by step. Section 9.1 discusses the ideas and principles for calculating approximate square roots. Section 9.2 deals with finding the approximate square root of any number between 100 and 10,000. Section 9.3 deals with finding the approximate square root of any number between 1 and 100. Section 9.4 deals with finding the approximate square root of any number between 10,000 and 1,000,000.

Section 9.1: Ideas and Principles for Calculating Approximate Square Roots (difficulty 5*)

Let's first look at **the idea of calculating the approximate square root**. Take numbers up to 10,000 as an example.

Firstly, we clarify what is the error of the estimation of the square root. We call T the true result of $\sqrt{Y}$ by calculator and we call X the estimate result of $\sqrt{Y}$ by mentally calculation. The error E is the absolute value of T – X which is expressed as E = |T – X|.

Secondly, our goal is to mentally estimate X closely to T, so make E as small as possible.

In this section, we can mentally calculate the square root of a number (Y) which is less than 10,000 step by step as follows to reduce the error E from 5 to 2.5 to 0.5 to 0.25.

For example, we mentally estimate $\sqrt{Y} = \sqrt{3415} \cong X$.

We know that the square root of any number between 1 and 10,000 is a number between 1 and 100. We only need to find the square value of these 100 numbers, and then find out which interval the number to be square rooted falls into, and we will know its approximate square root.

Through Part I of this book, we have learned to mentally calculate the squares of 1-100. As long as the number to be square rooted is compared with these 100 square values, the square root will be estimated with an error of less than 0.5. The disadvantage is that you have to calculate the squares of 100 numbers. Is there a way we can do it with fewer computational steps?

We will do this in four steps here.

Step 1: Divide 100 into 10 intervals, where each interval has 10 numbers. There are 11 endpoints. Calculate the square values of these 11 endpoints, and then find out which interval the number to be found for the square root falls into. See the example below.

X	0	10	20	30	40	50	60	70	80	90	100
X^2	0	100	400	900	1600	2500	3600	4900	6400	8100	10000

Because 2500 < 3415 < 3600, so 50^2 = 2500 < 3415 < 3600 = 60^2 so 50 < $\sqrt{3415}$ < 60. We can take the middle number 55 (X) from 50 to 60 as an approximation of $\sqrt{3415}$. $\sqrt{3415} \cong 55$ its error E = |T – X| < 5. We can validate it in the next sections.

Step 2: We divide the selected interval of 10 numbers into two equal intervals. Thus, the two intervals have 3 endpoints. As we know the square values of the two endpoints from step 1, we just need to calculate the square value of the middle point (the ones digit is 5) and then find out which interval the number to be found for the square root falls into. See the example below:

X	50	55	60
X²	2500	3025	3600

Note: We use Formula 1.1 to calculate $55^2 = 5 \times (5 + 1) * 100 + 5^2 = 3025$. See Section 1.2.

Formula 1.1: $A^2 = (10 * m + 5)^2 = m * (m + 1) \times 100 + 25$

Because it is an approximate calculation, we can also approximate **m * (m + 1) x 100** (50 x 60 = 3000). This is easy to calculate.

Because $3025 < 3415 < 3600$, so $55^2 = 3025 < 3415 < 3600 = 60^2$, so $55 < \sqrt{3415} < 60$. We can take the middle number from 55 to 60 as $\sqrt{3415}$, so the approximate value of $\sqrt{3415} \cong 57.5 (X)$, its error $E = |T - X| \leq 2.5$.

Step 3: We divide the selected interval (including the two endpoints with a total of 6 numbers) into 5 intervals. Each interval has 1 number, and there will be a total of 6 endpoints. Calculate the square value of the 4 endpoints respectively (from step 2 above, we know the square values of the two end points), and then find out which interval the number to be found for the square root falls into. We should have learned how to quickly calculate the square of any number up to 100 in Part I of this book. See the example below.

X	55	56	57	58	59	60
X²	3025	3136	3249	3364	3481	3600

Because $3364 < 3415 < 3481$, so $58^2 = 3364 < 3415 < 3481 = 59^2$, so $58 < \sqrt{3415} < 59$. We can take the middle number 58.5 from 58 to 59 as the approximate value of $\sqrt{3415}$. So $\sqrt{3415} \cong 58.5 (X)$. Its error $E = |T - X| < 0.5$.

Step 4: From step 3, we know the interval with two endpoints. We call the left endpoint as L and the right endpoint as L + 1. We divide this interval in the middle as two intervals in which the left interval between L and L + 0.5 and the right interval between L + 0.5 and L + 1. Please know that $L^2 + L = (L + 1)^2 - (L + 1)$. That means $L^2 + L$ is approximately the middle point of the interval of L and L + 1, or $(L + 0.5)^2 \cong L^2 + L$. So If $L^2 + L > Y$, the $L \le \sqrt{Y} \le L + 0.5$, that means $\sqrt{Y}$ is in the left interval. Otherwise $L + 0.5 \le \sqrt{Y} \le L + 1$ in the right interval. After we know which interval for $\sqrt{Y}$, we estimate X in the middle of the interval. The error is less than half interval (0.25). From this example, L = 58, $L^2 = 58^2 = 3364$, Y = 3415. So $L^2 + L = 3364 + 58 = 3422 > 3415 = Y$, so $58 \le X \le 58.5$, we estimate $X \cong 58.25$. Its error $E = |T - X| < 0.25$.

The interested readers can add the step 5 to reduce the error to 0.05. It is not too difficult. Hint: if $\sqrt{Y}$ is in the left interval, divide the left interval into new 5 smaller equal intervals. The square root of the distance of two endpoints of each interval is 0.1. Let L/5 = S, then divide the left interval from L^2 to $L^2 + L$ into (L^2, $L^2 + S$) ($L^2 + S$, $L^2 + 2S$), ($L^2 + 2S$, $L^2 + 3S$), ($L^2 + 3S$, $L^2 + 4S$), ($L^2 + 4S$, $L^2 + L$), then you can check which interval include Y. We estimate $\sqrt{Y}$ as the middle point of the interval. The error is < 0.05. You can work out how to estimate $\sqrt{Y}$ if $\sqrt{Y}$ is in the right interval. From the above example, S = 58 / 5 = 11.6. 3422 − 3415 = 7 < 11.6, so $58.4 \le X \le 58.5$. We estimate X = 58.45. However, I suggest most readers just learn till step 3, which is enough most of the time. After you understand the principle, you can do as much as you can.

Let's look again at **the principle of finding approximate square roots.**

We calculate the square root of a number y which is $\sqrt{y}$ (or called square root finding), which we can represent as, $x \le \sqrt{y} \le (x + e)$, or $x^2 \le y \le (x + e)^2$. Our purpose is to use square calculations to find the x that satisfies this condition, where the maximum error of $|\sqrt{y}|$ is e. If we estimate $\sqrt{y}$ in the middle point of the interval, we take $\sqrt{y} \cong x + e/2$, the error of $|\sqrt{y}| < e/2$. If e = 0, the square root is the exact value. For example: $\sqrt{3249} = 57$ because $57^2 = 3249$. $\sqrt{3415} \cong 58.44$ has no exact solution.

According to the above three-step calculations, as long as we can calculate the one-digit multiplications, we only need one step to find the square root under 10,000 with error < 5.

If we can also calculate the square with 5 in our minds, the error of the square root within 10,000 is less than 2.5 in two steps.

We learned the square speed algorithm (up to 100) in Part I of this book. We can thus use only three steps to find the square root of numbers up to 10,000 with the error < 0.5.

We express these 3 steps computations in a slightly formalized way.

Step 1: Find x_1 and make $x_1^2 \le y \le (x_1 + 10)^2$. So $x_1 \le \sqrt{y} \le (x_1 + 10)$, the maximum error of $\sqrt{y}$ is 10. if we take $\sqrt{y} \cong x_1 + 10/2 = x_1 + 5$, the error is $\sqrt{y} < 5$.

Step 2: Find x_2, make $x_2^2 \le y \le (x_2 + 5)^2$. So $x_2 \le \sqrt{y} \le (x_2 + 5)$, the maximum error of $\sqrt{y}$ is 5. If we take $\sqrt{y} \cong x_2 + 5/2 = x_2 + 2.5$, the error is $\sqrt{y} < 2.5$.

Step 3: Find x_3 and make $x_3^2 \le y \le (x_3 + 1)^2$. So $x_3 \le \sqrt{y} \le (x_3 + 1)$, the maximum error of $\sqrt{y}$ is 1. If we take $\sqrt{y} \cong x_3 + \frac{1}{2} = x_3 + 0.5$, the error of $\sqrt{y} < 0.5$.

If you wish, we can do two more steps as follows:

Step 4: Find x_4 and make $x_4^2 \le y \le (x_4 + 0.5)^2$. So $x_4 \le \sqrt{y} \le (x_4 + 0.5)$, the maximum error of $\sqrt{y}$ is 0.5. If we take $\sqrt{y} \cong x_4 + 0.25$, the error of $\sqrt{y} < 0.25$.

Step 5: Find x_5 and make $x_5^2 \le y \le (x_3 + 0.1)^2$. So $x_5 \le \sqrt{y} \le (x_5 + 0.1)$, the maximum error of $\sqrt{y}$ is 0.1. If we take $\sqrt{y} \cong x_5 + 0.1$, the error of $\sqrt{y} < 0.05$.

Section 9.2: Square Roots from 100 to 10,000 (difficulty 5*)

If $100 \le y \le 10000$, $10 = \sqrt{100} \le \sqrt{y} \le \sqrt{10000} = 100$, so $10 \le \sqrt{y} \le 100$. $\sqrt{y}$ can be any number from 10 to 100 (including decimals). We can represent it as, $x \le \sqrt{y} \le (x + e)$, or $x^2 \le y \le (x + e)^2$. Our task is to find x that satisfies this condition, so the maximum error of $\sqrt{y}$ is e. If we take $\sqrt{y} \cong x + e/2$, the error is less than e/2.

Let's look at a few examples first:
Example: $\sqrt{7569} = ?$

Step 1: Because $80^2 = 6400 < 7569 < 8100 = 90^2 = (80 + 10)^2$, so $80 < \sqrt{7569} < (80+10) = 90$. So, the maximum error e is 10. If we take $\sqrt{y} = \sqrt{7569} \cong x + e/2 = 80 + 10/2 = 85$, error < 10/2 = 5. In fact, $\sqrt{7569} = 87$. The error is 2.

Step 2: Because $85^2 = 7225 < 7569 < 8100 = 90^2$, so $85 < \sqrt{7569} < (85+5) = 90$. The maximum error e is 5. If we take $\sqrt{y} = \sqrt{7569} \cong x + e/2 = 85 + 5/2 = 87.5$, error < 5/2 = 2.5. In fact, $\sqrt{7569} = 87$. The error is 0.5.

Step 3: Because $87^2 = 7569 < 7744 = 88^2$, so $87 = \sqrt{7569} < (87+1) = 88$. The maximum error e is 1. If we take $\sqrt{y} = \sqrt{7569} \cong x + e/2 = 87 + 1/2 = 87.5$, error < 1/2 = 0.5. In fact, $\sqrt{y} = \sqrt{7569} = 87$. The error is 0.5.

Example: $\sqrt{9532}$ = ?

Step 1: Because 90^2 = 8100 < 9532 < 10000 = 100^2, so 90 < $\sqrt{9532}$ < (90+10) = 100. The maximum error e is 10. If we take = x + e/2 = 90 + 10/2 = 95, error < 10/2 = 5. In fact, $\sqrt{9532}$ = 97.63. The error is 2.63.

Step 2: Because 95^2 = 9025 < 9532 < 10000 = 100^2, so 95 < $\sqrt{9532}$ < (95+5= 100). The maximum error e is 5. If we take $\sqrt{y}$ = $\sqrt{9532}$ ≅ x + e/2 = 95 + 5/2 = 97.5, error < 5/2 = 2.5. In fact, $\sqrt{9532}$ = 97.63. The error is 0.13.

Step 3: Because 97^2 = 9409 < 9532 < 9604 = 98^2, so 97 < $\sqrt{9532}$ < (97+1) = 98. The maximum error e is 1. If we take $\sqrt{y}$ = $\sqrt{9532}$ ≅ = x + e/2 = 97 + 1/2 = 97.5, error < 1/2 = 0.5. In fact, $\sqrt{9532}$ = 97.63 The error is 0.13.

Example: $\sqrt{953}$ = ?

Step 1: Because 30^2 = 900 < 953 < 1600 = 40^2, so 30 < $\sqrt{953}$ < (30+10) = 40. The maximum error e is 10. If we take $\sqrt{y}$ = $\sqrt{953}$ ≅ x + e/2 = 30 + 10/2 = 35, error < 10/2 = 5. In fact, $\sqrt{953}$ = 30.87. The error is 4.03.

Step 2: Because 30^2 = 900 < 953 < 1225 = 35^2, so 30 < $\sqrt{953}$ < (30+5) = 35. The maximum error e is 5. If we take $\sqrt{y}$ = $\sqrt{953}$ ≅ x + e/2 = 30 + 5/2 = 32.5, error < 5/2 = 2.5. In fact, $\sqrt{953}$ = 30.87. The error is 1.63.

Step 3: Because 30^2 = 900 < 953 < 961 = 31^2, so 30 < $\sqrt{953}$ < (30+1) = 31. The maximum error e is 1. If we take $\sqrt{y}$ = $\sqrt{953}$ ≅ x + e/2 = 30 + 1/2 = 30.5, error < 1/2 = 0.5. In fact, $\sqrt{953}$ = 30.87. The error is 0.37.

Example: $\sqrt{321}$ = ?

Step 1: Because 10^2 = 100 < 321 < 400 = 20^2, so 10 < $\sqrt{321}$ < (10+10)= 20. The maximum error e is 10. If we take $\sqrt{y}$ = $\sqrt{321}$ ≅ x + e/2 = 10 + 10/2 = 15, error < 10/2 = 5. In fact, $\sqrt{321}$ = 17.92. The error is 2.92.

Step 2: Because 15^2 = 225 < 321 < 400 = 20^2, so 15 < $\sqrt{321}$ < (15+5)= 20. The maximum error e is 5. If we take $\sqrt{y}$ = $\sqrt{321}$ ≅ x + e/2 = 15 + 5/2 = 17.5, error < 5/2 = 2.5. In fact, $\sqrt{321}$ = 17.92. The error is 0.42.

Step 3: Because 17^2 = 289 < 321 < 324 = 18^2, so 17 < $\sqrt{321}$ < (17+1) = 18. The maximum error e is 1. If we take $\sqrt{y}$ = $\sqrt{321}$ ≅ x + e/2 = 17 + 1/2 = 17.5, error < 1/2 = 0.5. In fact, $\sqrt{321}$ = 17.92. The error is 0.42.

Exercise 9.2: In the exercises below, estimate the square root of each number within an error < 2.5. You can reduce the error to 0.5, or even 0.25.

$\sqrt{521}$ =? $\sqrt{2832}$ =? $\sqrt{3727}$ =? $\sqrt{4354}$ =? $\sqrt{5979}$ =?

$\sqrt{6621}$ =? $\sqrt{7789}$ =? $\sqrt{8249}$ =? $\sqrt{8964}$ =? $\sqrt{9767}$ =?

$\sqrt{721}$ =? $\sqrt{1832}$ =? $\sqrt{2727}$ =? $\sqrt{3354}$ =? $\sqrt{4979}$ =?

$\sqrt{5621}$ =? $\sqrt{6789}$ =? $\sqrt{7249}$ =? $\sqrt{7964}$ =? $\sqrt{8797}$ =?

Section 9.3: Square Roots from 1 to 100 (difficulty 5*)

If $1 < y < 100$, $1 = \sqrt{1} < \sqrt{y} < \sqrt{100} = 10$, $1 < \sqrt{y} < 10$. $\sqrt{y}$ can be any number from 1 to 10. Can we get the error down to 0.5, 0.25, or even 0.05? Absolutely we can do it.

The idea is to multiply each number by 100 first, and the interval we calculate goes from 100 to 10,000. We can then refer to the method in Section 9.2, and divide the result by 10, which is the final result. We formulate it as follows.

If $1 < y < 100$, $100 < 100y < 10000$. If $z = 100y$, $\sqrt{z} = 10\sqrt{y}$, $\sqrt{y} = \sqrt{z}/10$. We just have to calculate for the $\sqrt{z}$ (see Section 9.2), and we can get $\sqrt{y} = \sqrt{z}/10$.

Example: $\sqrt{95}$ = ?

$y = 95$, $z = 95 \times 100 = 9500$, $\sqrt{y} = \sqrt{95} = \sqrt{9500}/10 = \sqrt{z}/10$, so, we first calculate $\sqrt{z} = \sqrt{9500}$ =?

Step 1: $90^2 = 8100 < 9500 < 10000 = 100^2 = (90 + 10)^2$, so, $90 < \sqrt{9500} < (90 + 10)$. We choose $\sqrt{9500} \cong 90 + 10/2 = 95$, so the error of $|\sqrt{9500}| < 5$.

Step 2: $95^2 = 9025 < 9500 < 10000 = 100^2 = (95 + 5)^2$, so, $95 < \sqrt{9500} < (95 + 5)$. We choose $\sqrt{9500} \cong 95 + 5/2 = 97.5$, so the error of $|\sqrt{9500}| < 2.5$.

Step 3: $97^2 = 9409 < 9500 < 9604 = 98^2 = (97 + 1)^2$, so, $97 < \sqrt{9500} < (97 + 1)$. We choose $\sqrt{9500} \cong 97 + 1/2 = 97.5$, so the error of $|\sqrt{9500}| < 0.5$.

After three steps, we take $\sqrt{z} = \sqrt{9500} \cong 97.5$, error < 1/2 = 0.5. So $\sqrt{y} = \sqrt{95} = \sqrt{9500}/10 = \sqrt{z}/10 = 97.5 / 10 = 9.75$. The error < 0.5/10 = 0.05. The fact is $\sqrt{95} = 9.74679$, the error < 0.003.

Example: $\sqrt{66}$ = ?

y = 66, z = 66 x 100 = 6600, $\sqrt{y}$ = $\sqrt{66}$ = $\sqrt{6600}$/10 = $\sqrt{z}$/10, so, we first calculate $\sqrt{z}$ = $\sqrt{6600}$ = ?

Step 1: 80^2 = 6400 < 6600 < 8100 = 90^2 = $(80 + 10)^2$, so, 80 < $\sqrt{6600}$ < (80 + 10). We choose $\sqrt{6600} \cong$ 80 + 10/2 = 85, so the error of |$\sqrt{6600}$|< 5.

Step 2: 80^2 = 6400 < 6600 < 7225 = 85^2 = $(80 + 5)^2$, so, 80 < $\sqrt{6600}$ < (80 + 5). We choose $\sqrt{6600} \cong$ 80+ 5/2 = 82.5, so the error of |$\sqrt{6600}$|< 2.5.

Step 3: 81^2 = 6561 < 6600 < 6724 = 82^2 = $(81 + 1)^2$, so, 81 < $\sqrt{6600}$ < (81 + 1). We choose $\sqrt{6600} \cong$ 81+ 1/2 = 81.5, so the error of |$\sqrt{6600}$|< 0.5.

After three steps, we take $\sqrt{z}$ = $\sqrt{6600} \cong$ 81.5, $\sqrt{y}$ = $\sqrt{66}$ = $\sqrt{6600}$/10 = $\sqrt{z}$/10 = 81.5 /10 = 8.15. The error < 0.5/10 = 0.05. The fact is $\sqrt{66}$ = 8.124, the error < 0.026.

Example: $\sqrt{8}$ = ?

y = 8, z = 800, $\sqrt{y}$ = $\sqrt{8}$ = $\sqrt{800}$/10 = $\sqrt{z}$/10, so, we first calculate $\sqrt{z}$ = $\sqrt{800}$ = ?

Step 1: 20^2 = 400 < 800 < 900 = 30^2 = $(20 + 10)^2$, so, 20 < $\sqrt{800}$ < 30. We choose $\sqrt{800} \cong$ 20 + 10 / 2 = 25, so the error of |$\sqrt{800}$|< 5.

Step 2: 25^2 = 625 < 800 < 900 = 30^2 = $(25 + 5)^2$, so, 25 < $\sqrt{800}$ < 30. We choose $\sqrt{800} \cong$ 25 + 5 / 2 = 27.5, so the error of |$\sqrt{800}$|< 2.5.

Step 3: 28^2 = 784 < 800 < 841 = 29^2 = $(28 + 1)^2$, so, 28 < $\sqrt{800}$ < 29. We choose $\sqrt{800} \cong$ 28 + 1 / 2 = 28.5, so the error of |$\sqrt{800}$|< 0.5.

After three steps, we take $\sqrt{z}$ = $\sqrt{800} \cong$ x + e/2 = 28 + 1/2 = 28.5, error < 1/2 = 0.5. So $\sqrt{y}$ = $\sqrt{8}$ = $\sqrt{800}$/10 = $\sqrt{z}$/10 = 28.5 /10 = 2.85. The error < 0.5/10 = 0.05. The fact is $\sqrt{8}$ = 2.828, the error < 0.022.

Exercise 9.3: In the exercises below, estimate the square root of each number within an error < 0.25. You can reduce the error to 0.05.

$\sqrt{81}$ =?	$\sqrt{91}$ =?	$\sqrt{95}$ =?	$\sqrt{72}$ =?	$\sqrt{64}$ =?
$\sqrt{56}$ =?	$\sqrt{60}$ =?	$\sqrt{49}$ =?	$\sqrt{42}$ =?	$\sqrt{46}$ =?
$\sqrt{36}$ =?	$\sqrt{30}$ =?	$\sqrt{33}$ =?	$\sqrt{25}$ =?	$\sqrt{20}$ =?
$\sqrt{24}$ =?	$\sqrt{16}$ =?	$\sqrt{12}$ =?	$\sqrt{15}$ =?	$\sqrt{10}$ =?

Section 9.4: Square Roots from 10000 to 1000000 (difficulty 5*)

If $10000 < y < 1000000$, $100 = \sqrt{10000} < \sqrt{y} < \sqrt{1000000} = 1000$, $100 < \sqrt{y} < 1000$. $\sqrt{y}$ can be any number from 100 to 1000. Can we get the error down to 10, 5, or even 1? Let's try it out.

This idea is similar to Section 9.3, but by dividing each number by 100, the interval we calculate goes from 100 to 10000, and then referring to the method in Section 9.2, the result multiplied by 10 is the final result. The idea is just opposite of what we do in Section 9.3. We formulate it as follows:

If $10000 < y < 1000000$, $100 < y/100 < 10000$. If $z = y/100$, $\sqrt{z} = \sqrt{y}/10$, $\sqrt{y} = \sqrt{z} * 10$. As long as we work out first $\sqrt{z}$, we can get $\sqrt{y} = \sqrt{z} * 10$.

Example: $\sqrt{951234} = ?$
$y = 951234$, $z = y/100 = 951234/100 = 9512.34$, $\sqrt{z} = \sqrt{9512.34} = \sqrt{951234}/10$, so, we first calculate $\sqrt{z} = \sqrt{9512.34}$. Because of the approximation, we round off the number after the decimal point. So, we just ask for $\sqrt{z} = \sqrt{9512}$.

Step 1: $90^2 = 8100 < 9512 < 10000 = 100^2 = (90 + 10)^2$, so, $90 < \sqrt{9512} < 100$. We choose $\sqrt{9512} \cong 90 + 10/2 = 95$, so the error of $|\sqrt{9512}| < 5$.

Step 2: $95^2 = 9025 < 9512 < 10000 = 100^2 = (95 + 5)^2$, so, $95 < \sqrt{9512} < 100$. We choose $\sqrt{9512} \cong 95 + 5/2 = 97.5$, so the error of $|\sqrt{9512}| < 2.5$.

Step 3: $97^2 = 9409 < 9512 < 9604 = 98^2 = (97 + 1)^2$, so, $97 < \sqrt{9512} < 98$. We choose $\sqrt{9512} \cong 97 + 1/2 = 97.5$, so the error of $|\sqrt{9512}| < 0.5$.

After three steps, we take $\sqrt{z} = \sqrt{9512} \cong x + e/2 = 97 + 1/2 = 97.5$, error $< 1/2 = 0.5$. So $\sqrt{y} = \sqrt{951234} = \sqrt{9512} * 10 = \sqrt{z} * 10 = 97.5 * 10 = 975$. The error $< 0.5 \times 10 = 5$. The fact is $\sqrt{951234} = 975.31$. Error < 0.31.

Example: $\sqrt{662389} = ?$
$y = 662389$, $z = y/100 = 662389/100 = 6623.89$, $\sqrt{z} = \sqrt{6623.89} = \sqrt{662389}/10$, so, we first calculate $\sqrt{z} = \sqrt{6623.89}$? Because of the approximation, we round off the number after the decimal point. So, we just ask for $\sqrt{z} = \sqrt{6624}$.

Step 1: $80^2 = 6400 < 6624 < 8100 = 90^2 = (80 + 10)^2$, so, $80 < \sqrt{6624} < 90$. We choose $\sqrt{6624} \cong 80 + 10/2 = 85$, so the error of $|\sqrt{6624}| < 5$.

Step 2: $80^2 = 6400 < 6624 < 7225 = 85^2 = (80 + 5)^2$, so, $80 < \sqrt{6624} < 85$. We choose $\sqrt{6624} \cong 80 + 5/2 = 82.5$, so the error of $|\sqrt{6624}| < 2.5$.

Step 3: $81^2 = 6561 < 6624 < 6724 = 82^2 = (81 + 1)^2$, so, $81 < \sqrt{6624} < 82$. We choose $\sqrt{6624} \cong 81 + 1/2 = 81.5$, so the error of $|\sqrt{6624}| < 0.5$.

After three steps, we take $\sqrt{z} = \sqrt{6624} \cong x + e/2 = 81 + 1/2 = 81.5$, error $< 1/2 = 0.5$. So $\sqrt{y} = \sqrt{662389} = \sqrt{6623.89} * 10 = \sqrt{z} * 10 = 81.5 * 10 = 815$. The error $< 0.5 \times 10 = 5$. The fact is $\sqrt{662389} = 813.87$. Error < 1.13.

Example : $\sqrt{79597} = ?$

$y = 79597$, $z = y/100 = 79597/100 = 795.97$, $\sqrt{z} = \sqrt{795.97} = \sqrt{79597}/10$, so, we first calculate $\sqrt{z} = \sqrt{795.97}$? Because of the approximation, we round off the number after the decimal point. So, we just ask for $\sqrt{z} = \sqrt{796}$.

Step 1: $20^2 = 400 < 796 < 900 = 30^2 = (20 + 10)^2$, so, $20 < \sqrt{796} < 30$. We choose $\sqrt{796} \cong 20 + 10/2 = 25$, so the error of $|\sqrt{796}| < 5$.

Step 2: $25^2 = 625 < 796 < 900 = 30^2 = (25 + 5)^2$, so, $25 < \sqrt{796} < 30$. We choose $\sqrt{796} \cong 25 + 5/2 = 27.5$, so the error of $|\sqrt{796}| < 2.5$.

Step 3: $28^2 = 784 < 796 < 841 = 29^2 = (28 + 1)^2$, so, $28 < \sqrt{796} < 29$. We choose $\sqrt{796} \cong 28 + 1/2 = 28.5$, so the error of $|\sqrt{796}| < 0.5$.

After three steps, we take $\sqrt{z} = \sqrt{796} \cong x + e/2 = 28 + 1/2 = 28.5$, error $< 1/2 = 0.5$. So $\sqrt{y} = \sqrt{79597} = \sqrt{796} * 10 = \sqrt{z} * 10 = 28.5 \times 10 = 285$. The error $< 0.5 \times 10 = 5$. The fact is $\sqrt{79597} = 282.13$. Error is $|282.13 - 285| = 2.87 < 5$.

In order to reduce the error to be less than 0.5, we need to go two more steps as I introduced in **Section 9.2**.

Step 4: Based on the Step 3, $28^2 = 784 < 796 < 841 = 29^2$, we know $796 - 784 = 12 < 28$, so $\sqrt{796} < 28.5$, we choose $\sqrt{796} \cong 28 + 0.25 = 28.25$, so the error of $|\sqrt{796}| < 0.25$.

Step 5: $S = 28/5 = 5.6$. As $796 - 784 = 12 > 5.6 * 2 = 11.2$ and $12 < 5.6 * 3 = 16.8$, so $28.2 < |\sqrt{796}| < 28.3$. So we choose $\sqrt{796} \cong 28.2 + 0.05 = 28.25$, so the error of $|\sqrt{796}| < 0.05$.

Finally $\sqrt{y} = \sqrt{79597} = \sqrt{796} * 10 = \sqrt{z} * 10 = 28.25 \times 10 = 282.5$. The error $< 0.05 \times 10 = 0.5$. The fact is $\sqrt{79597} = 282.13$. Error is $|282.13 - 282.25| = 0.12 < 0.5$.

Exercise 9.4: In the following exercises, estimate the square root of each number with an error of no more than 5. You can reduce the error to 2.5, or even to 0.5 for some interested readers only.

$\sqrt{813251} = ?$ $\sqrt{918926} = ?$ $\sqrt{956723} = ?$ $\sqrt{722015} = ?$ $\sqrt{641324} = ?$

$\sqrt{56789} = ?$ $\sqrt{60123} = ?$ $\sqrt{49230} = ?$ $\sqrt{42134} = ?$ $\sqrt{46460} = ?$

CONCLUSION

Part I of this book introduces the formula for the rapid mental calculation of square of a number. By introducing the magic number, the complexity of the square calculation is greatly reduced, and we can mentally calculate the square value of a number within 1000. The square calculation within 100 is especially simple. The rapid calculation of the square of a number supports the rapid calculation of the magic multiplication in Part II and the rapid calculation of the magic number square root in Part III.

One of the core points of this book is:

With only two simple formulas, the square calculation of any two-digit number can be converted into the square calculation of a number less than 25 and the calculation of addition (or subtraction), so that the complexity of the square calculation is greatly reduced.

Formula 2.1 and **Formula 2.2** convert the square calculation using a small magic number m (< 25). It becomes the calculation of the square of the magic number m and the calculation of addition (or subtraction).

$$\textbf{Formula 2.1:}\ A^2 = (50 + m)^2 = (25 + m) \times 100 + m^2$$

where $A = 50 + m$ or $m = A - 50$. Since m is 50 less than A, the calculation of m^2 (where $-25 < m < 25$) is much easier than the calculation of A^2 (where $25 < A < 75$).

$$\textbf{Formula 2.2:}\ A^2 = (100 + m)^2 = (A + m) \times 100 + m^2$$

where $A = 100 + m$ or $m = A - 100$. Since m is 100 less than A, the calculation of m^2 (where $-25 < m < 25$) is much easier than the calculation of A^2 (where $75 < A < 125$).

Part II of the book presents a formula for rapid multiplication of any two digits.

Another core of this book is—

> The multiplication of any two digits is transformed into the calculation of squares and addition and subtraction by a simple formula. The calculation of squares has been simplified by the above Formulas 2-1 and 2-2, and the following Formula 6.1 has also been simplified.

$$\textbf{Formula 6.1:}\ A_1 * A_2 = (A_1 + m)^2 - m^2$$

where $m = (A_2 - A_1)/2$.

When $A_2 - A_1$ is odd, m is not an integer and it is more complicated to calculate m^2. We do the following transformations:

$$A_1 \cdot A_2 = A_1 \cdot (A_2 - 1 + 1) = A_1 \cdot (A_2 - 1) + A_1$$

When we use Formula 6.1 to calculate $A_1 \cdot (A_2 - 1)$, the magic number m is an integer.

Part III of the book introduces the square root rapid calculation. It is the outstanding contribution of this book. No other rapid calculation book has a quick algorithm that approximates the square root using mental calculation. The magic square root algorithm in this book is simple and easy to learn. The threshold to learn is so low that as long as you have elementary school arithmetic, you can learn it.

We can use a shape like capital Y to describe this book. This is the essence of this book.

The lower half of the Y is the foundation of this book, which is composed of the square rapid calculation of Part I.

The upper left of the Y is the multiplication algorithm (see Part II), which is supported by the lower part of Y. We can convert multiplication to squares to calculate the answer.

The upper right of the Y is the square root rapid calculation (see Part III), which is supported by the lower part of Y. Square roots are the inverse operation of squares. Use the square approximation method to find an approximation to the square root.

If you understand the essence of this book, you will soon be making new discoveries by yourself.

SELF-ASSESSMENT QUESTIONS

Below are the same 60 self-assessment questions. How many can you answer correctly now? How long does it take?

65 x 65 =	85 x 85 =	58 x 58 =	53 x 53 =
47 x 47 =	45 x 45 =	108 x 108 =	103 x 103 =
98 x 98 =	92 x 92 =	12 x 12 =	18 x 18 =
22 x 22 =	38 x 38 =	28 x 28 =	68 x 68 =
89 x 89 =	72 x 72 =	82 x 82 =	112 x 112 =
103 x 108 =	101 x 107 =	13 x 18 =	14 x 16 =
96 x 99 =	92 x 97 =	53 x 57 =	66 x 64 =
24 x 84 =	36 x 76 =	33 x 73 =	49 x 69 =
18 x 98 =	44 x 64 =	46 x 54 =	57 x 63 =
28 x 32 =	78 x 82 =	37 x 43 =	61 x 79 =
58 x 62 =	54 x 66 =	51 x 67 =	49 x 51 =
39 x 43 =	37 x 45 =	58 x 98 =	21 x 81 =
19 x 89 =	34 x 94 =	25 x 105 =	32 x 72 =
152 x 152 =	198 x 198 =	302 x 302 =	499 x 499 =
$\sqrt{92}$ =?	$\sqrt{35}$ =?	$\sqrt{8088}$ =?	$\sqrt{3028}$ =?

Name: _____ Time: _____ minutes: _____
Number of questions answered correctly: _____
Date: ___year___ month ___ day

How much have you improved since the first time you did this? Please do the test again in a week. If you make progress, applaud yourself!

CONCLUDING REMARKS

So that's it. As mentioned in the Forward, the focus of this book is not to make you memorize all the formulas or do all the exercises, but to cultivate your interest in mathematics, inspire you to think positively, and develop the ability to discover patterns. Has our goal been achieved?

The author was inspired by middle school teachers when he was a teenager and has continued to think and summarize for more than 50 years. Any discovery is not easy to come by, and the "Flash of Genius" is not accidental. Only positive thinking can trigger inspiration, and only perseverance can transform inspiration into results.

If you have any comments or suggestions on this book, or come up with a better and easier algorithm, you are welcome to contact me directly. When we reprint this book, your suggestions or algorithms will be incorporated to benefit more readers.

ABOUT THE AUTHOR

Professor Chengqi Zhang is an internationally renowned scholar on artificial intelligence. By the end of 2023, he and his collaborators have published around 400 articles in international academic journals and conferences and five academic monographs. He is a Distinguished Professor of Artificial Intelligence at the University of Technology Sydney, Australia, Pro Vice-Chancellor of the University of Technology Sydney, and Chairman of the Australian Computer Society National Committee for Artificial Intelligence. He is also the General Chair of the International Joint Conference on Artificial Intelligence (IJCAI)-2024, and the overseas appraisal expert of Chinese Academy of Sciences.

Professor Zhang graduated in 1982 from Fudan University with a degree in Computer Science, and has since obtained a Master's degree from Jilin University, a PhD degree from the University of Queensland and a Doctor of Science from Deakin University, all related to Artificial Intelligence. He was the first person from mainland China to publish a paper in the academic journal "*Artificial Intelligence*" in 1992.

www.ingramcontent.com/pod-product-compliance
Ingram Content Group UK Ltd.
Pitfield, Milton Keynes, MK11 3LW, UK
UKHW060050240426
12048UKWH00019B/1413